I0122693

Homo Schizo One

Alfred de Grazia

Homo Schizo One

Human and Cultural Hologenesis

Metron Publications

ISBN: 978-1-60377-087-3

Library of Congress Catalogue Number: 2014905381

Copyright ©Alfred de Grazia 1983, 2014

Second edition
Metron Publications, P.O. Box 1213

Princeton, NJ 08540-1213

To

Sebastian,

primus inter pares

Table of Contents

FOREWORD

Most scholars believe that man has progressed since his original appearance on earth. Probably so, but it has been a strange kind of progress, not well understood, and often showing a negative balance of the "bad" over the "good."

Some scholars believe that man is a rational animal. In limited ways he is, but, again, it is a strange kind of rationality, more ape-like than other traits of humans that are called "non-rational." For, to preview an argument that comes later, man is continually seeking ways to reestablish the uninterrupted instinctive responses of his forebears, and this is the homologue of "rationality." When Descartes wrote of animals as machines, he was obviously unaware that the precise "rationality" of man, which he, of all philosophers, elevated to awesome status, was just this homologue of the machine and animal.

So constrained and confused is whatever is called human rationality, that I prefer to call mankind by the name *homo schizo*, that is, *homo sapiens schizotypus*, rather than *homo sapiens*. Humans were created and are born schizotypical, with a set of traits to be distinguished in this book. They were from the first, and are now, more schizophrenic than otherwise. What is called "rational" is a derivation out of schizotypicality. This line of argument is also pursued in a companion volume, *Homo Schizo Two: Human Nature and Behavior*, which deals with today's people.

Here we are concerned with the evolution of mankind, a field densely covered with literature, but with many a sprouting mystery and contradiction that has resisted the spray of evolutionary formulas. The field is surprisingly vulnerable to a variety of pests, if iconoclastic views may be termed such. It invited questions. And to these I attempt answers.

By what means did hominid become man? By electrochemical means, and suddenly. Was the change large or small? The change was substantially minute, but profound in its consequences. When did it happen? Recently - about one thousand reproductive generations ago, which comes to about 260 memorial generations. What role did great natural forces play? They precipitated and perpetuated the change. Did culture spring up with, or did it lag behind, the human transformation? Culture sprang up with the gestalt of human creation. How many symptoms of mental illness are innate in man? All of them. How many cultures are "sick"? All of them, but the sickness is "normal." Can *homo schizo* aspire to become *homo sapiens?* One can aspire to a fiction, but cannot achieve it. Occasionally, a person, or even a group, can reach a delicate equilibrium, which can be called "reasonable," thus becoming *homo sapiens schizotypus.* Anything more than that is most uncertain.

The answers are tentative, as must be many scientific propositions. They may appear far-fetched, but rightly so, because they must be brought in from faraway fields. They would be more firm if only a few students of anthropology, linguistics, genetics, psychology, natural history, and early human behavior were disposed to drink deeply from their primeval fountain of self-doubt, and thereafter to re-examine their data.

I regret not being able to credit the full literature and cannot pretend to have slighted nobody. Especially am I concerned about the lurking work which may have quite escaped research, the work that would have bolstered my strained defenses or, for that matter, penetrated them, and which will emerge later, in a recapitulation of the Mendelian scenario. I recall that Mendel's genetic work "was published in 1865, in plenty of time for Darwin to amend his view in later editions of the *Origin,"* or so says Julian Huxley. His evolutionary theory badly needed the evidence of mutations in biology. Others, the same Julian Huxley for one, have made excuses for Darwin, and I hope that someone will do the same for me.

Alfred de Grazia

CHAPTER I

SLIPPERY LADDERS OF EVOLUTION

Scientists tracing the origins of man face an almost impossible task. So little remains of the beginnings that the very dirt around a suspected visit of early man is prized. They must grasp for anything tangible, a fossil bone, a chipped stone, a coprolite. Yet here we are, on the trail of man's most important original trait, self-awareness, an intangible phenomenon that cannot fossilize. Few, even today, would contradict what the geneticist, Ralph Gerard, said in 1959: "I don't think any of us has the remotest idea of why subjective awareness developed." [1]

Self-awareness is the consciousness of self. Practically every human, perhaps everyone, can stand off and look at himself. In fact, he does so normally, does so frequently, does so readily and at so early an age that maybe even the baby must think "I am I." He is self-conscious before he can speak. The physical boundaries of the self, fingers, toes, ears, nose and eyes are matters of interest to the infant who teaches them to himself in a matter of months. Fixing mental boundaries goes on endlessly. Probably he

---

[1] On p.188 of Volume III, *Issues in Evolution*, Sol Tax and Charles Callendar, eds., of *Evolution After Darwin*, Chicago: University of Chicago, 1960, hereafter cited as E.D.

begins the study of himself in utero, even though he must wait for his deathbed to conclude it.

Granted we cannot discover directly the appearance of self-consciousness in fossils, we may seek its concomitants. Anything denoting symbolism is a valid clue. Apes use sounds to convey moods, intent, and information; there is no use denying that this is symbolic behavior. So humans have to employ double abstraction to be different: the sign and signal, plus a reference that is not tangible, as for instance a wind, a direction, a ghost, an absent party, a glyph on a tree or rock, a burial, a sign of yesterday, a signal for tomorrow. But what should we do with the chimpanzee 'Congo, ' who dabbled in painting, turning out hundreds of compositions in a style typified by bunched and fanned brush strokes? [2]

A second valid clue to self-awareness is a tool. Sharpening a stone for use shows a sense of the design that may be inherent in a recalcitrant object, and is a valid indicator of human abstraction. Human-seeming animals are almost totally bereft of clubs, spears, pounders, drums, ropes. If they may grasp a twig and poke out ants from a hole, they cast it away when the hunt has ended. They do not improve it or look after it or burden themselves with it for very long.

Does walking on two feet, bipedalism, mark the advent of self-awareness? A baby is self-conscious before it can walk; but, no matter, the different traits need not appear in perfect succession. Congenitally crippled babies become human rapidly; again, the human setting fills the gap. That bipedalism may have preceded self-consciousness is easy to contemplate (perhaps because it is easier to 'sell out' self-awareness than a physical trait). But the mind balks at a four-footed self-conscious creature, even though babies are very human while still in the crawling stage. I think that we must admit that bipedalism may be a precursor or an invention but not a proof of self-awareness.

Fire-making is sometimes accredited as a sign of humanness. Fire may have 'always' been used. Birds and other animals, including primates, play about natural fires and eat roasted vegetable and animal matter consumed by the flames.[3] A natural fire may borrowed and preserved for a long time. But any group that could conserve fire was probably able to make it by friction, especially if in the habit of striking rocks together. The humanness

---

[2] D. Morris, "Primate's Aesthetics," 70 *Natural History* (1961), 22-9.

[3] E. V. Komarek, Sr., "Fire and the Ecology of Man," Tall Timbers Foundation, Tallahassee (Florida), March 1967, 151-3.

of fire-use depends, then, upon how it is obtained and whether it is preserved.

THE HUMAN BRAINCASE

Ultimately we would have to play a trump card: the large brain. Can we not assign the birth of self-awareness to the appearance of the first modern cranium. Thus, typically, a physical anthropologist such as Le Gros Clark will arrange the fossil cranial discoveries in order of time and size. The scale might begin with a chimpanzee of 300 to 600 cubic centimeters of cranial capacity, proceed to an australopithecine of from about 450 to 800 cc, up through homo erectus who might achieve 1280, then through homo neanderthal with an average higher than our own (1300-1610 cc), then back to modern man with 900 to 2300 cc - elapsed time being set at four million years. At what point of skull size does the hominid leave off and the human begin? It would beg the question to answer: "when tool-making is associated with the skull." John Buettner-Janusz says properly: "Unfortunately too much anthropological writing has focused on cranial volume when there is no evidence that a critical threshold for cranial volume need be exceeded for such 'higher' activities as tool-making and, by implication, culture."[4]

The conventional answer is that we do not know precisely, but that we can assume that the cerebrum, evolving with the size of the cranium, became ever more clever until it conceived of fire-making, tools, speech, and abstract non-entities. There are reasons to doubt this scenario. We have no place in this book for Julian Huxley's exuberant declaration, that evolution "simply is not just a theory any longer; it is a fact, like the fact that the earth goes around the sun and that the planets do all sorts of things."[5] Nor can we follow naively the theory that as with anatomy, so with culture: culture, too, evolves, as originally with Tylor, Spencer, and Morgan, and still now with many anthropologists.[6] However, we agree with these latter that Boaz and his followers were excessively wrought up to claim, as did B. Laufer, that "the theory of cultural evolution is… the most inane, pernicious, and sterile theory in the whole realm of science."[7]

A human brain consumes 20% of the energy resources utilized by the person as a whole. At the same time only 2 to 4% of the cerebrum is said to

---

[4] *Origins of Man*, N. Y.: Wiley, 1966.

[5] III *ED* 265, also 107.

[6] As with Marshall D. Shalins, Elman R. Service, eds., *Evolution and Culture*, with papers also by D. Kaplan, T. G. Harding, and Leslie A. White, Ann Arbor: University of Michigan, 1960.

[7] *Ibid, v.*

be activated, even at peak periods. Obviously there might be an energy crisis if we could work our brains very hard. One may suspect that the brain grew large without the 'intention' or 'specific purpose' of working, much less thinking.

This seems more plausible when we consider that the bilaterality of the cerebrum is not necessary. The human mind can function well with one hemisphere, if training and acculturation occurs on the basis of just the single hemisphere. Acquiring a single hemisphere would not be 'handicapping,' as would, say, a single eye or leg. The genetic instruction for a double cerebrum is part of the bilateral anatomy that reaches far out among the animal orders. Once again, we have a surplus; it is not persuasive to claim that a second hemisphere is 'good' to have upon the accidental loss of one hemisphere, and thereupon involve 'natural selection.'

Supposing, however, a single hemisphere and a 400 cc brain -- less than a third of the average human but one-half of the fast learning brain of the one-year-old baby or of homo erectus -- it would appear that, if this were functioning physiologically in a human way, it would be functioning behaviorally, too, in a human way. One would have, if nothing new were added along with size, the same mental and cultural abilities that we have at present. One would operate humanly with less than the brain capacity of australopithecus.

Dwarves with well-proportioned bodies of 2 1/ 2 ft in height, and with brains weighing one-third (14 ounces) of the ordinary human's brain may be sometimes stupid, but they speak fluently. A high adult I. Q. on the Stanford-Binet intelligence test "is possible with about one-third of the total cerebrum lacking." But "adaptive" intelligence suffers at less than the 30% level. So says one authority.[8] He was perhaps unaware that, at about the same time as he was writing, a hydrocephalic Englishman, with one-tenth of the normal cerebral volume (10%) was doing well socially and in his university studies.

Another disturbing thought occurs: the weight of brain of the australopithecus was probably heavier, in proportion to his body size, than that of the modern human. This would support the idea that australopithecus should have been as clever as ourselves, or conversely, we might well be more stupid than australopithecus, if it were not for - what? Putting aside the unconvincing though popular view that, point-by-point, evolving man grew in brain size and in adaptive control of the environment, an argument that is part biological and part cultural but in

---

[8] Ward C. Halstead, "Brain and Intelligence," in L. A. Jeffress, ed. *Cerebral Mechanisms in Behavior,* N. Y.: Wiley, 1951, 251.

both cases implausible for reasons stated elsewhere, the source of the difference between the stupid hominid (assuming such was the case for the forebear of australopithecus) and the clever human must rest in a specialization of the brain and/or in its electro-chemical state and operations. My opinion here - and in the accompanying volume - is that both types of change occurred: specialization and a new electro-chemistry.

F. M. Bergounioux is persuaded that intelligence "is a phenomenon with no connection whatever with the physiological structure that supports it." So it seems, and one can observe hovering in his unusual essay the ultimate resort to teleological creationism such as Teilhard de Chardin developed.[9] The theory of homo sapiens schizotypus may, however, bridge this chasm between the subtlest human behavior and the physiological housing. Something other than brain growth was responsible for humanization.

THE SEARCH FOR A BETTER APE

A book on human origins written in the last century presents the same basic ideas as a book lately published; there is little new of importance in the recent book. The main difference is that about 1900, Mendelian mutations, actual changes in the germ plasma, were accepted by many geneticists as the main factor in the alteration of species. Although it could have been used to rehabilitate catastrophism, this discovery was used to reinforce the shaky foundations of the dominant Darwinian evolutionism.

Whereas the old book asked only modest amounts of time for the human race to develop from the ape, the new book asks for up to five million years. Aided and abetted by modern 'time-telling' techniques, such as the potassium-argon test, the new book can fit many skull-cases, jaw-bones and some extremities that have been uncovered into a long-time frame. Many comparative studies have been made of primates and people, showing, for example, how they walk or what relationship their blood hemoglobin contains. But no old evolutionist ever doubted the cousinship of man and ape: go to the zoo and see for yourself.

Evolutionary theories have to venture in fine detail into what came first. For evolution is uniformitarian, gradual, compounded bit by bit. Thus, a ladder of culture has been assembled. First, crude stone pounders and cutters, then use of fire, then many other developments, partly anatomical and partly cultural: cannibalism, walking upright, right-handedness, premature parturition, improved weaponry, crinkled brains, deft digitry,

---

[9] "Notes on the Mentality of Primitive Man," in S. L. Washburn, ed., *Social Life of Early Man*, Chicago: Aldine, 1- 64, 11.

weak dentition, improved diet, signalling, thinking ahead, fortifications, speech, burial of the dead, and so on.[10]

Many disputes have arisen as to priorities among the numerous steps forward in social evolution; no two ladders have the same rungs. If one were to collect a shelf of all major works on human evolution since and including the work of Charles Darwin, and took from each its 'first, ' 'truly human, ' 'necessary, ' 'all-important' steps, and then examined the list, he would feel bemused: each author builds his own ladder; each 'new' trait is the crucial trait that set off man from the ape. Sometimes the rungs are anatomical, at other times cultural; they may also be geological -- events of the rocks, ice, climates, the geomagnetic field, or of geochronometry.

An interesting ladder-scheme, unfortunately not well-developed, is offered by Walter Garre and called *The Psychotic Animal: A Psychiatrist's Study of Human Delusion*.[11] He believes that man, in evolving anatomically over millions of years, developed more and more tools and artifacts. Man was proud of his abilities and became, indeed, increasingly megalomaniac. He began to seek goals in the sky and on earth that he could not possibly obtain until finally he went mad. Vanity, then, is the nemesis of man, and the therapy for the human psychosis is to reconcile man to what is possible. Garre's ladder is amusing and at least more logical than most; his theory is, however, very lightly constructed.

To Freud, writing in 1930, "the upright posture of man was the start of his fateful development."[12] By getting his nose off the ground and putting his genitals up front, man exhibited himself and felt shame. Further, man could never gratify his sexual drive fully and therefore had to seek all kinds of sublimation, "all the cultural developments that are summed up by the word sublimation." Presumably this set of events would have preceded the events that gave him the truly human oedipal complex, the day when he and his brothers killed the old bull father in order to possess sexually the females, and felt ever thereafter an intensification of guilt[13] - or, to avoid implying that Freud contradicted himself, the great guilt as against the small shame.

In another widely read and more respected treatise, J. Bronowski stresses the development of omnivorous eating habits before other traits,

---

[10] *Cf. inter alia* J. N. Spuhler, et al., *The Evolution of Man's Capacity for Culture*, Detroit: Wayne U., 1959, chap. I., and S. L. Washburn and R. Moore, *Ape into Man,* Boston: Little, Brown, 1973.

[11] N. Y.: Human Sciences Press, 1976.

[12] *Civilization and Its Discontent*, 1930, NY: W. W. Norton, 1950, 46.

[13] *Totem and Taboo*, 1913, trans. 1950, N. Y.: W. W. Norton, 140ff.

beginning with australopithecus and moving through Neanderthal to modern man:

> The consequences for the evolution of man were far- reaching. He had more time free, and could spend it in more indirect ways, to get food from sources (such as large animals) which could not be tackled by hungry brute force. Evidently that helped to promote (by natural selection) the tendency of all primates to interpose an internal delay in the brain between stimulus and response, until it developed into the full human ability to postpone the gratification of desire.[14] Thus Bronowski momentarily sighted the instinct-delay, but was diverted into adding a rung to the ladder.

Dozens of carpenters and ladders are in the race. But each author has his detractors, who say such things as: 'You cannot eat meat without cooking it, ' or 'You can cook but still not be reflective, ' or 'Lower animals were omnivorous first.' The most effective way yet found to handle the disputatious crowd is to give everyone time - one, five, even ten million years. Then every ladder can climb to the same lofty level of modern humans who can do everything. What I propose here may be more effective: remove the ladder and let everyone in through the front door; they are all right, at the same time!

LEGENDS OF CREATION

'Let everyone in - do you mean even the creationists?' I am not so sure, but let us make a case for the legendary accounts of human origins. It is not impossible to do so. Man has no memory of being a hominid, much less an ape. He insists, however, that he remembers being created. The ready conclusion - one which has been proposed from the earliest times - is that mankind was humanized abruptly. This event was universally depicted in theological language as a divine creation. Hence scientists of the past century, in ridding themselves of religious constraints, ceased to consider whether, even without divine intervention, humanization might have occurred in a natural quantavolution. Charles Darwin, to begin with, did not attend, when his disciple, Thomas Huxley, wrote him in 1860 not to be too rigid with the adage, "Nature makes no leap" (natura non facit saltum). Darwin repeatedly termed the adage a "canon."

---

[14] *The Ascent of Man*, Boston: Little, Brown, 1973, 44-5.

In the historical record from its beginnings, and in the treasured oral records of non-literate peoples of today, mankind is portrayed as a divinely created being. He was fashioned, by beings of a higher order. Homo schizo apparently knew long before Aristotle that an effect had to have a sufficient cause. We may be curious as to why they did not claim eternity, why they did not accept the idea of a world beyond time, why they postulated a chaos followed by a creation. Nor did the earliest cosmologists venture that humans were descended from the lower animals, as much as they may have lived among and respected animals. Yet scholars commonly argue that clever primeval men invented their divine makers because they were not clever enough to imagine how they might otherwise come to exist upon the earth.

Peoples of all types of culture insist, with a unanimity that deafens modern scholars, that they were created, not evolved.[15] The Hopi Indians say that after the world was spun out and nicely formed and enlivened with plants and animals, twin gods made people and gave them speech and wisdom. The Wyot Indians maintain that the first people were furry and talked badly; a universal deluge was visited upon them, and a brother-husband and sister-wife brought forth the good new people.

The Eskimo Creator elicited people out of a scattering of seal bones. The Quiché Mayans proposed that twin gods filled the great void with water and earth; living creatures were made, but their voices could not praise specifically their creators. Whereupon mankind was made of clay, and the clay melted, requiring another attempt. "At first, it spoke, but had no mind." Abandoning clay, the gods resorted to wood. These wooden creatures could not walk properly, nor did they worship their creators. They were annihilated in hurricanes and deluges of black rain. The monkeys are their survivors. Now the gods made fine men, out of corn, so fine that the gods had to cast a mist before their eyes to prevent their knowing too much; and later the gods made them wives who came to them in their sleep.

The Swahili of East Africa adopted Islamic creation theory, which goes back to Judaic theory, which has man created from clay, which is also the Christian belief. One pygmy group of Zaire has god creating an 'Adam and Eve' and punishing them for violating his commandment, and a second story of the god creating humans as fruit of a special tree of life. The god of the Ngombe of Zaire let his human creations live with him in the sky. Then he exiled a troublesome woman with her son and daughter to earth, and

---

[15] A number of the cases comes from Barbara C. Sproul, *Primal Myths: Creating the World*, N. Y.: Harper and Row, 1979.

from these came the human race. (But a hairy stranger also mated with the daughter and their offspring brought evil and sorrow to the world.)

To the ancient Mexicans it seemed that the first race of men, created by one of the gods out of ashes, was destroyed by jealous gods in a flood, and the people became fish. Other ages intervened before the present one, "the Fifth Sun." In the fourth age the people were "ape-men" *(tlacaozomatin)*. In the fifth age, a god searched the regions of the dead for the bones of a couple of humans. These were found, ground up, and watered by blood from the penis of Quetzalcoatl. Now man, creature of divine self-sacrifice, must sacrifice continuously to keep the world in orderly motion.

Chinese legend has Nü-kua making people of yellow earth patties. Iranian Bundahism recites that man and bull were fashioned of the soil, and that the seed of life, made from the sky's light, was planted in their bodies. Various Greek nations claimed that the earth gave birth to their ancestors; for instance, the Thebans were born from the dragon's teeth sown by Cadmus. A Sumerian story conveys that Enki, the great god, ordered Mami, the mother goddess, to mix clay with the blood and flesh of a lesser god killed by the other gods. So it was done. As usual, the earth was thriving beforehand. And so it was when Elohim created Adam and Eve, the former out of clay, the latter out of a rib of Adam. The Egyptians believed man to be divinely fashioned of clay, too.

In Plato's dialogue, *Timaeus,* a didactic myth presents the faultless creator Demiurge, using the planets, including Earth, as factory sites, making human souls out of less pure materials than that of which the universe is made; and then "he distributed them, assigning each soul to its several star."[16]

The Skidi Pawnee of the Great Plains recited, "Our people were made by the stars; when the time comes for all things to end our people will turn into small stars and will fly to the South Star where they belong."[17]

But clay seems to be a favored material: "made of common clay." So also says Ovid, at the beginning of this era, but he adds "maybe." His Metamorphoses tells many a gruesome tale of people turning into monsters at the will of the gods, nor can we dismiss the idea that Ovid may have been trying to recount times of great radiation and mutation.[18]

---

[16] Giorgio de Santillana and Hertha von Dechend, *Hamlet's Mill,* Boston: Gambit, 1969, 306.

[17] *Ibid.,* 309.

[18] *Ibid.,* 252, 118.

MEMORIAL GENERATIONS

What could in fact the ancients remember, if anything? Oral traditions can survive for exceedingly long periods, at least some thousands of years. In the case of modern isolated tribes, and even in the case of the Hebrew and Indo-European Sumerian tradition, what reason do we give for our confidence that these stories cannot go back to the first stories of the first 'time-factored, ' that is, remembering or historical, mankind? Can any force change the roots of a myth? Through how many memorial generations of man do the roots of myth penetrate?

The statistical reports of groups exhumed from cemeteries and analyzed for age show average ages of death below 40 until recent times, but also persons who lived to advanced ages. (In a Bushman people numbering 248, living as marginally constrained hunter-gathers, 8% were from 60 to 80 years old).[19] If one memorial generation is the age difference between an old oral historian and a young child of a tribe, it may average fifty years. Ten thousand years gives only 200 careful sacred recitations; twenty thousand years gives 400. If all the peoples of the world pay sacred respects to what amounts to a story of the sudden appearance of humanity, this fact would seem to support the idea of a continuous story from the beginning of man.

Suppose that a psychologist and anthropologist, supported generously by the U. S. National Science Foundation and Institutes of Health, were to set up a chain of 800 story-tellers, sixty-year olds alternating with ten-year olds, and told the first person in the chain the Eskimo creation story. Would the 800th person repeat the essential story, granting such changes as 'seal bones' becoming bones of another animal? Let an awesome authority warn that the story must be retold with perfect accuracy, "lest you die."

A much more sophisticated study design is possible; my purpose here is to position the problem for intuitive comprehension. There are grounds for believing that a basic legend can go back even 100,000 years, an age conventionally assigned to homo sapiens, if it conveys a fundamental truth.

If the story goes back that far, or even if it does not, how does it happen that fine legends are not spun about the evolution of man from the animals? Or of his eternal existence? With ages of religious prejudice behind us, we must of course be contemptuous of descent from lower animals. Yet can we believe that the earliest men had to invent gods because they were so disgusted with their similarities to animals? Even when men lived close to animals, endowed them with human characters, and worshiped them as totems? And, too, the earliest stories and depictions around the world reveal,

---

[19] John E. Pfeiffer, *The Emergence of Man*, N. Y.: Harper and Row, 1972, 391.

for instance, bulls and women in sacred copulation, not to mention snakes and swans. T. Dobzhansky is therefore probably reasoning ad hoc when he says: "Infinity is a notion which most people find hard to conceive of. Creation myths were accordingly constructed to show that man and the universe did have a beginning."[20]

The thrust of legends, when scientifically considered, is directed at humanization as a discrete kind of event, remembered by a mind that recalls not what happened beforehand to itself but what happened then and ever thereafter -- a new kind of memory. And, we guess, this was and remained a fearfully composed memory, compulsively and obsessively recollecting itself. Somehow a barrier was suddenly thrust up between humans and animals.

Hans Bellamy alludes to the "remarkable fact that the mythologist, though he knows an immense number of creation myths, cannot point to a single one whose report starts right at the beginning of things... Almost everywhere we find the ordering of a chaotic muddle of pre-existing things, a formation or a reformation on an improved plan, a recreation rather than a creation in the primary sense of the term."[21] The Earth is fashioned out of the body of a vanquished monster, or fished out of the primordial sea, or created by the word of a demiurge, this last a favorite of later priests, so that, for instance, the creator gods assembled, and called "Earth!" and the Earth arose from the waters. As St. John said, "In the beginning was the word; the word pervaded God; the word was God." Afterwards man was created, as earlier stated. 'Of course, ' it can be argued, 'these are typical schizophrenic delusions, having no basis in reality. 'Very well -- although it is rather early in the book to accept our thesis that man was born schizophrenic and has always been schizotypical. Can we not also suggest here that man was striving in manifold ways to recall a hologenesis of mind and culture? And that he must have been a true human at the time of the events at issue?

It is in this connection, too, that we can address the extensive work of Mircea Eliade on *The Myth of the Eternal Return.* [22] For he finds everywhere in the world, and displaced onto all of the functions of life, such as farming and sex, a compulsion to conduct anniversaries and rites to commemorate the first great days of human existence, insisting that 'this is the way things were in the beginning, ' *illo tempore.* Eliade does not analyze the causes of this universal human behavior; he rests with the facts, uncovered with so much

---

[20] *Mankind Evolving,* 1962, N. Y.: Bantam, 1970, 1-2.

[21] *Moons, Myths and Man,* London: Faber and Faber, 1936, 165.

[22] Princeton University Press, 1964.

toil. Here we take what seems to be the necessary step beyond, asserting that humans may remember their origins.

Now, if this is so, then the cultural, or 'intrinsic', memory of man must be extremely long, or the time allocated to human origins must be far too long. Probably the moment has not yet arrived for calling into question the estimates of the duration of human becoming. We still have not heard the stories -- we shall not call them legend -- told by the scientists who have worked with the rocks, the bones, and the artifacts composing the underground history of mankind.

NATURAL SELECTION

Doubts about the efficacy of a ladder of evolution begin with questions about the means of constructing the ladder, that is, the machine of natural selection. Charles Darwin titled his influential work *The Origin of Species by Natural Selection.* Although his mentor, the geologist Charles Lyell, had employed the word "evolution" since 1832, Darwin did not use the term in his own book that came 27 years later. An "unfolding" of new traits was certainly implied, in biology as in geology, especially since Darwin thought (rather vaguely, it seems) that new traits emerged from within individuals as they competed for survival within their species and with representatives of other species.

On the other hand, Darwin used the term "natural selection" 414 times, and "selected" or "selection" an additional hundred times. The heavy employment of the term suggests that he was using it not only as a referent, but also as an active substitute for real natural operations and in place of non-existent evidence.

In general, darwinism has provided a century of confused thought about natural selection. Looking back from today, it is difficult to understand how the idea could so have captured the minds of scientists, granted that its public appeal was large. We should not forget that Darwin (and Wallace, whose ideas on natural selection paralleled his own) received the idea behind natural selection upon reading Malthus who in turn was keen on justifying the laissez-faire notion of a struggle for survival in economic affairs. He demonstrated persuasively that, while the means of subsistence were growing arithmetically, population was growing by geometrical progression, with an ultimate resolution only through famine, disease, and war. It is surprising that even the marxists, who were so suspicious of bourgeois ideology, should have overlooked the import of this connection, when adopting the idea of evolution by natural selection. Marx did associate

Darwinism with liberal English economics, but did not insist upon following through the consequences of his surmise.

One may allude to Darwin's inattention to Gregor Mendel's studies of plant genetics. Why on the other hand, would he have taken the first opportunity to put down Mivart's work (1871), which argued that evolution could only be explained as a series of saltations.[23] It seems that Darwin was bent upon taking his inspiration from a hard-headed economic realist rather than from other biologists, perhaps only to guard his idea of natural selection, but perhaps also because he realized that sudden leaps in evolution would, when it came to the journey from ape to man, open the door once more to the religious creationists.

Most cases advanced to illustrate the concept of natural selection turn out to be Lamarckian environmentalism or question-begging. The pattern was set by Darwin himself. He was even capable of statements "that mutilations occasionally produce an inherited effect."[24] More recently, we have Washburn and Howell declaring that "it was altered selection pressures of the new technical-social life which gave the brain its peculiar size and form."[25] Elsewhere, Washburn has it that, "In a very real sense, tools created homo sapiens."[26] So Buettner-Janusz, claiming that culture put severe demands upon the brain, causing it to evolve.[27]

That is, man is a kind of self-fulfilling prophecy, governing his own evolution in some of its most critical aspects such as brain size and specialized brain areas, arguments that verge beyond the Lamarckian toward several other hazy theories on the fringes of scientific discussion - teleological explanations, inherent Platonic forms seeking their realization, etc. Where does all this evolutionary sap come from that now causes the mind to burgeon and then again fashions the tool for the mind to use? But such has been a common form of arguing around the weakness of natural selection in its stark logical definition.

More often, natural selection is proven by a kind of question-begging. Thus, a trait of a species, one not found in a fossil relative, is given an ex post facto justification by natural selection. A common formulation reduces to this: a species which did whatever was done tended to survive in greater numbers. But no proof is offered. Both natural selection and mutation

---

[23] Ernst Mayr, "The Emergence of Evolutionary Novelties," I. *ED.* 354ff; St. George Mivart, *Genesis of Species,* London: Macmillan, 1871.

[24] *Descent of Man,* 1871, 1883, 440, *cf.* 435.

[25] "Human Evolution and Culture," II *ED* 52.

[26] Spuhler, *op. cit.,* p31.

[27] *Op. cit.,* 352.

theory abound with the stated or implied premise that whatever changed must have changed because the change helped the species to survive.

A typical problem occurs with asymmetrical brain organization in the human, which accompanies, but not necessarily in a mutually causative relation, handedness – right-handedness in about 87% of the species. Left-handed people are more brain-bilateral, both anatomically and functionally. Their left and right crania exhibit less asymmetry and their speech areas are less centralized in their dominant hemisphere.

There occur thereupon the typical rationalizations of brain asymmetry and handedness: these 'help the species to survive by promoting dexterity; ' and 'the left hemisphere, with an accomplished right hand, can carry out its dominating wishes and calculations. '

In acute brain lesions of the dominant hemisphere, left-handed persons suffer less speech loss than right-handed persons. "If the majority of the LH (approximately 70%) have bilateral representation of speech, this atypical organization would spare them from the more severe and prolonged effects of a unilateral lesion that would be seen in the RH person whose speech mechanisms are more laterally differentiated." [28] Now, if enough clubs smashed enough skulls in the billions of fights during the ascent of man, and if speech were important after the battles ended, and if other variables were not present, then man should by now be left-handed and retrogressed to bilaterality.

However, apart from these particular 'if's,' there occur scores of additional 'iffy' variables. For instances, left-handers are considered wrongheaded by most people, and maybe inferior, so might they not be exterminated? Also, might not left-handed club-wielders be more surprising and effective in battle and therefore reduce the right-handers with evolutionarily significant frequency? Or be employed by right-handers to fight and disproportionately die, while the right-handers remained home to breed?

And might not the right-handers, being more asymmetrical, be also more schizoid, and being more schizoid, be more paranoid, assertive and socially dominant over the left-handers; but schizotypicality is fostered, too, by invidious cultural discrimination, so should not the left-handers like Leonardo da Vinci more than hold their own in the evolution of the species. So do we not have a statistical stand-off, what evolutionists might gratefully refer to as 'an evolutionary equilibrium of 70 and 30 proportions resulting from the operations of natural selection'? This line of thought could go on

---

[28] Paul Satz, "A Test of Some Models of Hemispheric Speech Organization in the Left- and Right- Handed," 203 *Science*, 16 March 1979, 1133.

almost indefinitely, with every question begged by the interposition of the magical term "natural selection."

GRADUALISM

Charles Darwin felt committed to the view that man must have arisen from lower primate forms to his present eminence by a ladder of incremental changes. In *The Descent of Man,* he conceived of "a series of forms graduating insensibly from some ape- like creature to man as he now exists" so that "it would be impossible to fix on any definite point when the term 'man' ought to be used." [29] (He used the terms "gradations" and "gradual" some sixty times in *The Origin of Species.*)

The history of fossil anthropology has seen many attempts to prove Darwin's insensible gradations to be the correct scenario for human development. Thus, a century later, Le Gros Clark, the authoritative physical anthropologist referred to earlier, thought "it is evident that a closely graded morphological series linked *Australopithecus* through *homo erectus* with our own species *homo sapiens.*" [30]

A prominent zoologist, Ernst Mayr, could in 1951 set forth a fine case for cultural elaboration being attendant upon brain enlargement. [31] A decade later he might say the same of all speciation, but only by leaving out careful considerations of time, of the mathematics of permutations and combinations, of the earliest actual origin of the rich intra-species gene pool being called upon the allow remarkable adaptation, and by skirting the edges of Lamarckian environmentalism even while denying it. [32]

In considering the advent of homo sapiens, alert scepticism about the language of natural selection and mutation theory will send many a popular view crashing to the ground. There is little in the known history of human evolution that can be called upon to show that natural selection, adaptation, the survival of the fittest, or even 'mutation as an aid to natural selection, ' has played any part in the present constitution of mankind. But, to question-begging, evolutionary discourse adds a ping-pong game in which a frustrated

---

[29] Page 541.

[30] *The Antecedents of Man,* Chicago: Quadrangle, 1971, 359.

[31] "Taxonomic Categories in Fossil Hominids," 15 Cold Spring Harbor *Symposium on Quantitative Biology* (1951), 109-17.

[32] I. *ED.* 354ff.

natural selection explanation bats the ball to mutation theory, which, frustrated in turn, bats the ball back to natural selection.

Moreover, the same scepticism may be indulged regarding the mania for extending time backwards to great lengths. A theory of natural selection, plus point-by-point mutation, plus an unchanging or very slowly changing natural environment are going to require very much time to effect the multitude of alterations distinguishing the human being from its imagined primate archetype. The ladder of evolution has to be very long.

However, we may not use the long ladder to prove that time is long, even though time must have been long in order to build such a ladder. Time has to be proven long by independent criteria and tests. The scientific world has conveniently forgotten that Darwin conceived of natural selection as having originated and developed all species of life to their present state within a time span which, by present standards that move toward two or more billion years, would make of him a rapid evolutionist. Relative to a small span of time, the years allocable to the ascent of man were negligible by contemporary guesses; even then time was short, no doubt explaining some of the exasperation of gentlemen of the day, who could feel the hot apish breath of their ancestors on the back of their necks.

The ideology still prevails, suffusing the field of study with three hypotheses: that one fossil form has progressed to another very gradually, that the elapsed time has been long, and that the culture traits have budded upon the branches of anatomical changes. But also (see Washburn, above) the brain can bud on the branches of culture; thus, tools excite brain growth.

What are we allowed to think of the evidence if we disrobe our minds of the ideology of darwinism for a moment? Humanoid types have been dispersed over most of the Earth. Different types lived at the same time and even in the same places. There are no provably transitional types. Stone tools and artificial dwellings have characterized the earliest bipedal large-brained types. "Stone tools are *prima facie* evidence that there was sufficient neurological material for culture."[33] But can culture (that is, humanization) be potentiated for three or more million years without realizing a breakthrough somewhere? Can the measures of time be wrong? With all this, must we not begin to consider whether there occurred some quantavolution, some saltation, as opposed to a gradual evolution?

Must we take a position on the duration of humanizing evolution in order to develop the theory of homo schizo? Suppose that we accept a 5-million-year evolution from hominidal ancestors to modern man. Can we then say that man has changed bit by bit over this period of time and very

[33] Buettner-Janusz, *op. cit.,* 349.

gradually became the schizoid type that we know today? And, to address C. Darwin, could we then speculate that, at some point near the end of this period, this changing anatomy finally produced an outburst of cerebration and culture?

Also, did man lose his instinctive behavior bit by bit, with blunting and delay occurring in one after another case, until finally he became modern? Was he, incipiently, and then more and more, self-aware and was he more and more frightened and anxious as time went on, until finally he achieved full self-consciousness?

If so, what brought on this gradual change? Was it a series of mutations, all leading in the same direction ('directed evolution') or a continuous process of natural selection breeding a creature more effective at survival? But it is not possible for mutations to work so rapidly under present and recent natural conditions. Nor, considering how many changes would be required and that these changes had to be transferred in a set of successive 'chain reactions' to the species wherever its habitat, has there been time for natural selection.

SEVERE LIMITS TO NATURAL SELECTION

And what is natural selection? We come back to the question. Darwin complains, "I cannot... understand how it is that Mr. [Alfred] Wallace maintains, that 'natural selection could only have endowed the savage with a brain a little superior to that of an ape. '"[34] It may be that natural selection, if it makes sense at all, is capable only of ensuring survival. The fittest may survive, but to be 'fittest' means only fitter than the next individual of one's species, and being a member of a species that is reproductively fitter than whatever species at the moment may be cutting into this reproductivity. Natural selection is a measure of the influence, at a given moment, of a life form. It is the interaction of life forms and their living and inorganic environment favors the genetic descent of certain forms and the extinction of others, whether of the same or of different species. From this, it is logical that an individual life form that is favored tends to expand in numbers.

But if the environment at Time 'X' changes erratically or quantavolutes, then the changes within an individual and species that have occurred up to Tx can promptly lose their merits as factors in natural selection. What helps for survival this year may hurt survival next year. So it is that natural

---

[34] *Descent of Man*, 432.

selection is a more persuasive idea if one is a uniformitarian, believing processes in nature have always been as they are now.

Persuasive it may be, but still not statistically probable. As soon as all the variables are emplaced in the correlation matrix, the likelihood of natural selection collapses. For, what uniformitarian evolution provides in the way of infinite chances of 'advance' must be provided as infinite chances to 'retreat, ' hence infinite contradictions. The general reliability of natural selection in producing an 'advance' must be close to zero.

The environment which effects species selection is so changeable even under uniformitarian conditions that no 'line of evolution' can be credible as an effect of natural selection. One moment a virus, the next a drought, the next an elimination of a competing species by other causes than direct competition, then a chance mutation then a hundred other selective forces play upon the situation of a species. And, of course, the holistic structure and function of an organism, where thousands of interdependencies interact with each ongoing moment, are utterly beyond the selective capacities of nature, as these are presently construed. And, if one flees to time for protection, they are quite beyond the capabilities of the longest time.

When a gathering was convoked at the University of Chicago in 1959 to celebrate a hundred years of *On the Origin of Species by Natural Selection,* and after much wisdom was spoken and the final discussions ensued, there occurred within minutes a blurting of confessions and hopes.[35] Ernst Mayr was concerned with evolutionary outbursts along some lines after many millions of years of stability, and wondered how so many extinctions occurred, considering "the extreme sensitivity of natural selection, doing the most incredible and impossible things." Emerson said that he himself was of the opinion that "We need much more precise information on the evolutionary time dimension within all the biological sciences - behavior and development and so on," and A. J. Nicholson regretted that whereas much attention had been given to the disappearance of unfit forms, little attention had been given to the "replacement of unfit forms."

Such research specifications have, needless to say, gone unfulfilled for another twenty years. David Raup ventured to say that "we have even fewer examples of evolutionary transition than we had in Darwin's time,"[36] and a conference held in 1981 at his institution, the Field Museum, in Chicago,

---

[35] III. *ED.* 141-2. *Cf.* Steven M. Stanley, *The New Evolutionary Timetable,* N. Y.: Basic Books, 1981; Francis Hitching, *The Neck of the Giraffe,* N. Y.: Mentor, 1982; T. M. Schopf, ed., *Models in Paleontology,* San Francisco: Freeman, 1972.

[36] "Conflicts between Darwin and Paleontology," quoted by L. R. Godfrey in *Natural History,* June 1981, 9.

focused entirely upon the possibility of macroevolutionary periods, without facing squarely the non-uniformitarian mechanisms that might have produced them, such as catastrophes.[37]

I shall not argue that a busy god exists: but I would point out that hard-headed materialists of the evolutionist camp, who are quick to cite the human stupidity which can treasure a religious delusion for thousands of years, should not have trouble in recognizing that they, too, have been laboring under a delusion, that of natural selection, for 150 years. God is not the only ideological delusion making the rounds of humanity.

If modern man has taken a long time to evolve and if the changes were on the ladder, say, of ramapithecus - australopithecus - pithecanthropus - homo, there should have occurred a great many intermediate types, each with some distinctly 'progressive' concatenation of bones and behavior. These have been claimed; they had to be claimed. But, as we shall see, the known types are several at most. Also, it is unlikely that more than one or two additional types will be found.

Generally, the prevailing modes of thought act to suppress this kind of observation, and let presumptuous expressions such as that of Le Gros Clark pass without serious criticism. As evidenced by the Piltdown Man fraud, whenever a missing link or transitional type seems to emerge, it is eagerly seized upon.[38] In any event, should not such types have survived, even the several known fossil hominids? Up to the present, man has not been able to exterminate his primate relatives, and presumably the hominids would have been more clever and elusive than the apes and monkeys.

Very recently (May 2, 1981) a commentator in the *New Scientist* could sloganize the controversy as 'lucky survivors' *versus* natural selection. Species do not arise by any provable natural selection but only on occasion flourish thereby or decline, and even then almost always by happenstance that has practically nothing to do with "survival of the fittest" as a selective mechanism. Mutation is the seemingly general mode of creating new species and perhaps of destroying many, but then mutation is another matter, an electro-chemical event offering advantageous or disadvantageous possibilities in a given environment. Many a 'hopelessly inept species' lives on and there are many 'marvelously adapted' fossils of extinct species. Millions fewer of extinct fossil forms are found than 'should be found, ' if one is to judge by the number of existing species.

Exponential reproducibility is a *prima facie* case versus the refined general theory of natural selection. Natural selection by any means whatsoever,

---

[37] Roger Lewin, "Evolutionary Theory Under Fire," 210 *Science* Nov. 1980, 883-7.

[38] J. S. Weiner et al., *The Piltdown Forgery*, London: Oxford, 1955; *Nature*, 2 Nov. 1978.

except general catastrophe, reduces to its largest component, exponential reproducibility. Clever little wings, a nose that sniffs better, and all the thousands of alterations of species and individuals designed as 'improvements by natural selection, ' are as nothing compared with the formidable propensity of every species to reproduce in infinite numbers.

Seen in this light, the fact that should be astonishing, but seems to impress few, that the simplest virus or bacterium survives as well or better than the most complex species, can only mean that catastrophe and reproducibility determine natural selection. For the rest, natural selection has been a fol-de-rol, diverting developmental biology from more important business. Darwin prepared an epitaph for his main concept when, in expounding gradualism, he predicted, "so will natural selection, if it be a true principle, banish the belief of a continued creation of new organic beings, or of any great and sudden modification in their structure."

"WAVES OF EVOLUTION"

Scholars generally believe that four waves of evolution have occurred in the ascent of man. The first was of pro-human apes, all fossils now, such as Aegyptopithecus, Dryopithecus, and Ramapithecus, who inhabited Old World locations from 34 to 8 million years ago (so it is said). "There are, in fact, no ape fossils from anywhere after about eight million," notes Johanson.[39] These extinct beasts were without sign of human culture despite a fairly large brain. That they could have behaved in 'stupid' human ways or could have had descendants, also extinct, that might have done so, is not impossible. Adrian Desmond [40] illustrates well how modern apes are hovering upon the brink of self-awareness and of varied deliberate activities. Such intimations of humanity, which may be enhanced by future paleontological discoveries and modern experiments, are in line with our general theory here, as they are with conventional evolution. The mechanics of humanization, to be discussed in the next chapter, may have altered primate behavior in the same directions of ego-fracture and or delayed instinct response as they did in ourselves.

The second wave was australopithecine. Estimates of their age vary up to a million years in the case of individual finds and extend from a half-

[39] Donald Johanson and Maitland Edey, *Lucy,* New York: Simon and Shuster, 1981, 363. Washburn and Moore, *op. cit.* Buettner-Janusz, *op. cit.;* NY Times, Feb. 7, 1980 on new Aegyptopithecus discoveries by Elwyn Simone.

[40] *The Ape's Reflexion,* N. Y.: Dial, 1979.

million to several million years within the group of finds. Some 243 to 285 of these hominids are represented in fossil discoveries in Africa and Asia. The most famous come from Olduvai Gorge near Nairobi and the Afar Depression (" Lucy"). Some were discovered earlier and others are being uncovered. The brain of australopithecus could achieve 800 cubic centimeters, especially large in view of his small size; his ratio of brain to body bulk was greater that of modern man, 1/ 42 as opposed to 1/ 47 by one calculation. [41] His neck was proportionally longer too. He was completely adapted to bipedalism.[42] He was right-handed. His physique varied from "gracile" to "robust;" he weighed perhaps 32 to 39 kilograms, and resembled in musculature a modern Bushman of the same area.[43]

The third wave was pithecanthropus or homo erectus, who also spread out over Africa and Asia. He is found so close to australopithecus in certain excavations, as at Olduvai Gorge, that he probably lived at the same time. The most famous is Peking man from China. His brain attained 1200 cc., large also in relation to his stature. His time is guessed at anywhere from 100,000 to millions of years (or this whole range of time).

Other finds of homo erectus are adjudged in the same range. Homer Rainey reports Johanson's estimates of 3 to 4 million years for the Afar Depression homo of 1975 an 2 to 6 million years for the R. Leakey rift finds of 1972 and says that "several manlike and other Homo species were contemporary in very ancient times. Moreover they were toolmakers."[44] Soviet excavators at Azhch (near Erivan) have discovered remains, tools, and incised bear skulls, dated at 450,000 years.

Then came the proto-homo sapiens, who differ little from modern homo sapiens in anatomy. Often they are called homo erectus, with little reason save their arguable old ages. I doubt that the earliest of these would be considered non-human if their age were unknown. There came, too, the Neanderthal (316 specimen individuals) who was long considered sub-human until discovered co-habitating with our kind in Palestine. He is now given homo sapiens status, but not quite admitted to the club of homo sapiens sapiens. By then, and even before then, modern types were flourishing, so that some 400,000 years is an arguable age of full man in current anthropological circles.

---

[41] Buettner-Janusz, 146, 350-1, *et passim.*

[42] C. O. Lovejoy, "The Locomotor Skeleton of Basal Pleistocene Hominids," IX *Proceedings*, Congress, UISPP, 14 Sept. 1976, 157.

[43] Alan Mann, "Australopithecine Demography," *Ibid.* 181.

[44] Encyclopedia Britannica Yearbook, 1976, 260.

There are three main cultural periods to attach to these four waves. All of the creatures except the pro-human apes have worked tools, the most tangible signs of a culture. The Paleolithic is divided unsurprisingly into Lower, Middle, and Upper, the Lower going back to the earliest tools, which may be anywhere from 500,000 to 5m/ y old by conventional reckoning; in geological time this would be Middle Pleistocene to Pliocene.

After describing the habitual bi-pedalism of australopithecus, Wolpoff points out that the canine teeth of australopithecus do not differ significantly from those of homo erectus. He then describes the tool kit of australopithecus, saying, "Indeed, some of the australopithecine industries are surprisingly advanced. The Sterkfontein and Natron industries have been called Acheulian."[45]

Alberto Blanc helped rehabilitate Neanderthal man, accrediting him with ritual mutilation of skulls going back 250,000 years, in a style close to that employed in Bronze Age Germany and present-day mutilation practices in Borneo and Melanesia. Further, he pointed out that homo erectus (Peking man) was available in fragments of forty individual skulls; only one piece was entirely missing from all forty, the base or foramen magnum, signifying probable mutilation, and therefore a possible connection running all the way from homo erectus through Neanderthal to modern man.

> The reconstructed skull of Sinanthropus offers, therefore, an astonishing resemblance to the mutilated skulls of the "early" and "late" Neanderthals and to the skulls mutilated for the purpose of practicing ritual cannibalism in the Bronze Age of Germany and by the present head-hunters from Borneo and New Guinea.[46]

It is also probable that ritual skull mutilation signifies ritual cannibalism. He mentions the famous figure, "obviously the figure of the god or genius of the hunting people," of the Cave des Trois-Frères in Ariège, with the horns of a deer, paws of a bear, eyes of an owl, and tail of a wolf or horse. There is no reason to doubt his word that "the constant complexity of human beliefs is valid and abundantly proved, at least since the Upper Paleolithic."[47]

---

[45] Milford H. Wolpoff, "Competitive Exclusion Among Lower Pleistocene Hominids: The Single Species Hypothesis," 6 *Man* 4 (1971), 606.

[46] "Some Evidence for the Ideologies of Early Man," in S. C. Washburn, ed., *Social Life of Early Man*, 133.

[47] *Ibid.* 121.

F. Bordes, among others, lumps together the Lower and Middle Paleolithic, does not find them in America, and attributes to the long period an Acheulian and a Mousterian style. But he speaks of overlapping: "Prehistory is now at a point where we have to accept the idea of contemporaneity not only of different culture variants, but also of different cultures, and this not only in different provinces, but also in interstratification in the same region."[48] Acheulian and Mousterian have been noted to overlap, by Mellars and others. The Mousterian culture is also found in connection with Aurignacian Upper Paleolithic remains. The same type of person made both types of artifacts, or two types of people made both, thus being equally human.

J. E. Weckler writes, "it is no longer possible to maintain the idea that biface cores were the work of homo sapiens and flake tools the product of Neanderthal; for we know that generally in the Europe-Africa-India range the Levallois flakes and biface cores were made by one and the same people as parts of unified cultural assemblies."[49]

The Upper Paleolithic and Mesolithic are joined, too, in America as well as in the rest of the world. A report from Russia carries a shoe-print of an Upper Paleolithic hunter with evidence that the type worse trousers.[50] The modern races are probably present in the Upper Paleolithic. Australians go back now 100,000 years, according to a 1980 news report. Further, australoid types have been found in South Africa and Ecuador. North American Amer-Indian types have been pushed back into the Upper Paleolithic. The major Asian, Sinese or Mongolian types are on hand, and the Caucasians are amply present in the Mediterranean and Europe. Neanderthal probably merged with the caucasoids, rumors of extermination to the contrary notwithstanding. If the rock drawings of the Sahara and Southwest Africa are Upper Paleolithic, as their style might indicate, would their artists be negroid or caucasian, or mixed assemblages of types? The answer is still unknown, but that they were religious is undoubted.

Little time is required for human types to diffuse around the world. As if to confirm this conjecture, a recent dispatch carries the claim of Alan Thorne of Australian National University to have discovered fossil remains

---

[48] "Chronology of Paleolithic Cultures in France," in Renfrew, ed., *The Explanation of Culture Change: Models in Prehistory,* Pittsburgh, U. of Pitt., 1973. F. Ameghino, in several works at the turn of the century, claimed an Acheulian culture of the Lower Paleolithic in South America"

[49] "The Relationships between Neanderthal Man and *Homo Sapiens,*" 56 *Amer. Anthro.* (1954) 1011.

[50] Peter Kolosimo, *Spaceships in Prehistory,* Secaucus, N. J.: University Books, 1979, source not cited.

of Chinese humans in North Australia which date to at least 10,000 years.[51] That humans, ecumenically cultured, split off in early natural disasters, and that a land platform prevailed until about 6000 years ago during which they might move around in the Southeast Pacific, is considered in this book and in *Chaos and Creation*.

J. D. Birdsell thought Australia might have been settled within 720 years by pioneering negritos from Timor but places the date at 32,000 years ago, which I must regard as too long a time. He guessed that the australopithecines moved thousands of miles from South Africa to Southeast Asia in 23,000 years. This, too, seemed swift to him and to others: "Pleistocene man when spreading into unoccupied territory could have saturated it to carrying capacity... in amazingly short elapsed time."[52]

Yet Americanists long believed that men crossing the frozen arctic Bering Straits reached practically to Antarctica in 12,000 years. Now man is thought to be older in the Americas. I would maintain that man is as old in the Americas as anywhere else, but in any event his velocity of diffusion was much greater everywhere. No hominid or homo need have more than a few centuries to stretch around the globe. And, if hominids and homo were contemporary, and especially if all were "human," the occupation of the world by mankind need have consumed no more than a thousand years. (I would maintain this whether the world was land-covered -- see my Chaos and Creation -- or fragmented.) Furthermore, present racial differences are such as may have occurred in brief periods of isolation, followed by bursts of regional expansion of new types. The mechanism of such quantavolutions in the hominid sphere, as in the biosphere generally, is quantavolution in the natural sphere, catastrophes such as I depicted in *Chaos and Creation*.

The Neolithic period brought practically everybody everywhere to the stage where most people still are, except for some use of metal now in many parts around the world. Pottery, farming, domestication of animals, religion and many other cultural features are present everywhere. Yet, nowhere, strangely, is it claimed that the Neolithic is more than a few thousand years old, six to twelve thousand being the normal estimated range.

We need not consider this Neolithic Period here. No hominid or proto-homo-sapiens emerges during it. Also, as indicated above, nothing basically important seems to have distinguished the Upper Paleolithic from the Mesolithic. So far as human development is concerned, the cultural level of

---

[51] "Chinese 'First to Australia, '" *Melbourne Sun,* Aug. 14, 1982.

[52] "Some Population Problems involving Pleistocene Man," 22 *Cold Spring Harbor Symposium of Quant. Biol.* 1957, 67-8.

the Upper Paleolithic approaches that of the Neolithic (later on, I shall offer my evidence to this point). So the temporal question is whether homo schizo originated then, or in the Middle or Lower Paleolithic, bearing in mind that by Lower Paleolithic we must mean Early Pleistocene, with this period in turn moving back into what was once thought to be Pliocene, and perhaps even into the so-called Cretaceous.

The time problem is tied in with the manner of genesis. Did this human being originate in steps or by quantavolution, that is, all at once? Did his culture originate promptly with his physical origins, that is, hologenetically? In answering these questions, we shall be solving the problem of time. A quantavolution of human genetics and culture implies human hologenesis, and both imply a collapse of time scales. If timescales are deprived of anthropological, archeological, and legendary support, they must subsist upon geology and geochemistry. And if they cannot do so, they must be radically adjusted.

CHAPTER II

HOMINIDS IN HOLOGENESIS

Might all types of known hominids and proto-humans have been of the species homo sapiens (schizotypus) in physiology and culture? Might these and all modern races have appeared during the past 14,000 years? Might man have originated hologenetically in the holocene period, by quantavolution? Such is the line of questioning and argument to be followed here; outrageous as it may be to conventional theory, it may be also productive.

We have already noted that australopithecus had certain human qualities. We can pick up the analysis again. He was adequately supplied with cranial matter. Specimens exceeding the minimal brain size known for

normal humans have been discovered. His brain-body build proportions were modern. His size was that of many millions of modern people. His dentition was close to modern man's, far removed from the apes. He was bi-pedal and held his head high (higher than we do, said Louis Leakey). He was social. He used tools. He built enclosures. He was right-handed. It appears that his brain was hemispherically asymmetric, which introduces additional human potentials. McKinley, Wolpoff says, "demonstrated that Australopithecus (gracile and robust) followed a 'human' model of short birth spacing," and Mann showed that "the rate of australopithecine development and maturation were delayed, as in modern man, rather than accelerated, as in modern chimpanzees." (Based upon the timing of molar eruption.) There are no signs yet of his having had speech, but no evidence to the contrary; Louis Leakey thought he had a human palate. There have been few indications yet of his having been religious and artistic. There are signs of his having used fire.

He was connected with homo erectus in time and with the Acheulian-Chellean culture at Olduvai, which culture extends into the Terrafine of North Africa and is found also at Swanscombe and Steinheim, with practically modern man.

Opposing the theory that australopithecus was human stands largely the thesis that he is anatomically too different from modern man. To the forgoing response may be added the following: we do not know what are the limits of variation within the single species or how the principal distinction employed - that interbreeding be impossible - would apply here. It is of significance that Johanson had persistent doubts about classifying his fossil hominid, "Lucy." He argued that she might be called homo, but relented at the prospect, then, that all australopithecines would logically have to be regarded as of the homo line. Where would we go to find our hominid ancestors? The search for the missing link would begin again.

While at the University of Chicago, Charles Oxnard compared fossil australopithecines with living apes and men by fine measurements of the foot, pelvis, fingers and other bones, transferring the measures to computer tapes for multivariate analysis.[53] "Geometrically this is the equivalent of constructing and viewing from one position a three-dimensional model of the swarms (of points measuring similar objects) and then rotating and viewing the model from a new position that best separates the swarms." His studies suggest that the australopithecine bones are uniquely different from both man and the chimpanzee and gorilla.

Applying stress analysis to the bones supports his comparisons derived from the computer analysis in that the finger bones of man are incompetent for both knuckle-walking and hanging-climbing, whereas those of the Olduvai australopithecines are poor for knuckle-walking, but adapted for hanging-climbing. Oxnard believes also that australopithecus might have been better equipped to run than to stride bipedally. One wonders when the Olduvai "creature of the savannahs" stopped walking on his knuckles, how he used his hanging-climbing faculty, and why his hands were not scuppered for scooping fish from the successive "Lakes of Olduvai." Kamala, the Indian wolf-girl, went on all fours and could not stand, until after years of

---

[53] *Uniqueness and Diversity in Human Evolution*, Chicago, U. of C. Press, 1976, 169 *et passim*.

coaching; her hands were described as very strong and rough; she could run, rising on her digits. Her anatomy was normal for homo sapiens. The races of mankind are distinguishable as skeletons, but are one species; so the "hanging-climbing" hands of Olduvai man may be of minor importance if he were otherwise human.

Australopithecus may be a branch of the human line that habitually clung and climbed. Better yet, he may have maintained an ancestral feature that was finally bred out. It may be suggested, also, that he originated wherever homo emerged, and that the quantavolution was so open-ended as to provide a remarkable diversity of human types in the beginning, followed by a diffusion of these types around the world. Soon the types would double back, and merge with, or exterminate each other.

There is much diversity among the australopithecines themselves to fuel controversy; several attempts have been made to call new specimens by new species names. Generally, anthropologists have wanted to join them together as a single species, if only to avoid barren disputation and to make it easier to sum up the primordial situation in textbooks.

Charles Oxnard points out that a recent finding at East Rudolph, by Richard Leakey, of the keystone of a foot arch (talus) has been dated as the same age as, or older than, the Olduvai australopithecine. Yet the new find is much larger, more similar to that of modern men, according to the multivariate analysis of Bernard Wood. Richard Leakey found also a skull dated from two to three million years of age with an endocranial volume of 800 cubic centimers (the australopithecine volume being generally much less), showing an overlapping of cranial capacities with homo erectus. Then, too, an arm bone fragment from Kanapoi, dated at four million years, "has already been shown... to be very similar to that of modern man." In this phrase, "very similar," we can read within the range of variation of modern man.

The various pieces of evidence, according to Oxnard, add up to meaning that perhaps as long as 5,000,000 years ago (and the possibility is not lost that future finds may place this further back in time) there may well have been creatures living that were generally similar to homo erectus and therefore classifiable as man in a way that we must deny to any australopithecine (whether named *H. habilis, H. africanus* or whatever else)." That is, we should say, erectus was even more modern in anatomy than australopithecus. But probably present anatomical differences between the pygmies of the Congo and their tall black neighbors are as great as between australopithecus and homo erectus; they discourse in the same language (pygmies adapt and use neighboring languages), intermarry, have continuous commercial dealings, and in fact are symbiotic.

HOMO ERECTUS

Now it is homo erectus who comes to mind, and we would like to know whether he, too, might be human. If so, how did he come to be created? And when was he created? And if five million years old, or three, or one, or one-half, or one-tenth -- to cite various estimates -- why then should he have evolved so slowly until the Upper Paleolithic, which is variously reckoned at from 50,000 to 10,000 years ago?

Homo erectus cannot be dismissed from the motley ranks of modern man. He had large supraorbital ridges but so have some modern individuals and so, too, Neanderthal man, who had a cranium larger than modern man and a culture. Home erectus had a low skull, yet possessed the cranial capacity of the smaller skulls found among ourselves. And amongst ourselves, cranial size has little or no relationship to average intelligence and competence, or perhaps even to extreme intelligence.

Le Gros Clark, seeking to prove gradual evolution, wrote that by "the end of the Middle Pleistocene, the hominid skull had attained a degree of development very similar to modern man; indeed, except for the rather strongly developed supraorbital ridges, some of the cranial remains of this date are hardly to be distinguished from modern man." [54] His "Middle Pleistocene" was about 500,000 years ago, and in cultural terms might now be termed Lower Paleolithic, because well-developed stone age cultures dated at 250,000 years ago have been uncovered, as in Tadshik, U. S. S. R. Björn Kurten alludes to modern humans with a brain case of 1400 cc and using fire, discovered in Hungary at Varteszöllös in the 1960's and dated at 400,000 years ago. [55] Probably homo erectus and homo sapiens were contemporaries. Inasmuch as fire making was also assigned to Peking man, homo erectus whether 400,000 or four million years ago, it would seem that humans have been allowed an inordinately long time to sit around fires in a mental funk.

In 1975, field work at Koobi Fora in northern Kenya resulted in the demonstration of contemporaneity between KNM-ER 3733, an unequivocal homo erectus cranium, and KNM-ER 406, an obvious robust australopithecine. This was dramatic confirmation of earlier

---

[54] *Op. cit.*, 608.

[55] R.S. David, *et al.*, "Early Man in Soviet Central Asia," *Sci. Amer.*, 130-7; Björn Kürten, p.113.

interpretations that suggested the existence of two distinct hominid lineages in the African early Pleistocene.[56]

In Java, homo erectus and meganthropus were living side by side in the Middle Pleistocene.[57]

Presently, radiometric dating, particularly the Potassium-Argon test, is determining the ages of hominids, and this test is applied ordinarily to volcanic issue. The stretching of the time of hominids has gone on regardless of definitions of boundaries, and little attention is given to traditional geochronology. If the volcanic ashes imbedding a bone are adjudged to be two million years old, that is usually the end of age reckoning. So the hominids have gone back beyond the Pleistocene well into the Pliocene.

How baffling the time element can be is suggested in an incident. A skull of homo erectus was discovered in Kenya by Bernard Ngeneo, working under Richard Leakey. It was dated at 1.5 million years. Peking man, a prototype of homo erectus had been dated by non-radiometric methods at 0.5 million years or less. Leakey said, "this raises questions about the true age of Peking Man. The Chinese must develop a new, different way to date their sites for more accuracy. Upon re-examination, they'll probably find these fossils to be a million years older than now dated."[58]

In effect, the 40K-40A dating method is giving very old and by implication "good" results, and should be the sole method of plotting man's ascent! If so, some dates of hominid and *homo* fossils that were estimated before radiometric methods were employed may be useless. Or else these types lived for millions of years on Earth. As I stated earlier, modern types are now being found aged in the millions of years, not only skulls of modern volume but also modern bones, and now modern footprints.

---

[56] D. C. Johanson and T. D. White, "A Systematic Assessment of Early African Hominids," 203 *Science* (26 Jan. 1979), 326.

[57] R. Sartonon, "The Javanese Pleistocene Hominids," *Proceedings*, IX Congress UISPP, 14 Sept. 1976, 462, using fluorine tests.

[58] *New York Times*, March 9, 1976, 14, news conference.

PEKING MAN

Sinanthropus, the Chinese version of homo erectus, from Choukoutien, [59] probably had a cerebral mechanism for speech. He was also righthanded as judged by cerebral asymmetry and the way he made and used tools. His occlusal trough was the same as ours and he chewed the same way. His two lateral upper incisors "display a crown morphology quite typical for this region in various races of modern man." The upper central incisors were longer than in the Northern Chinese today. The lower molars were of a "generalized and progressive type... one whose slight modification in a given direction may readily produce a condition dominant in modern hominids" (The experts who say this make a comment that should be borne in mind when comparing ancient and modern man: that the most distinctive peculiarities of modern man are "degenerative in origin.")

Sinanthropus built fires and made artifacts of quartz; layers of ashes were uncovered and thousands of pieces of worked quartz. We will treat this matter when we discuss cultural hologenesis, but it may be worthwhile to mention here that the "Choukoutien formation must be considered as a perfectly homogeneous and distinct stratigraphical unit." To our view, this signals the possibility that the Choukoutien scenario was brief, not enduring for a hundred centuries or a thousand centuries.

An archaelogical columnar section illustrates the distribution of prehistoric culture in relation to deposits of North China, as known to Black and his collaborators half a century ago. I have tabulated it here. Note how crowded the holocene period is in relation to the Pleistocene and Pliocene sections, and yet how heavy its cultural development. So much time is allotted to the earlier periods because convention so dictates, i.e., such is the ruling paradigm of evolutionary time. But inspection of the contents of the column reveals plainly that practically all of its material could have been deposited in weeks, years, or centuries. The deposits are precisely of the type that occur in floods and storms: sandy lacustrine deposits, loams, loess, and gravel. (In *The Lately Tortured Earth*, I examine evidence of an extraterrestrial origin of the loess.) It is unlikely that hundreds of thousands of years elapsed, as the report declares. This idea is especially poignant because the Choukoutien fossils and artifacts were found in lenses of deposits that were swept into a rock cleft, fissure, or large cave, filling it up, until, in our day, they were come upon in the course of quarrying.

---

[59] Davidson Black, Teilhard de Chardin, C. C. Young, W. C. Pei and Wong Wen Hao, *Fossil Man in China*, Series A, Noll, Geological Memoirs, Geological Survey of China, Peiping, May 1933, repr., AMS Press, New York, 1973.

That the total setting is recent is attested to by occasional unsuspecting sentences in the reports: the fissure contains such a wide range of fauna from Late Pliocene and Upper Pleistocene (at least 1 myr) "that it is not easy to decide to which of them it stands more closely related," so it is placed as Lower Pleistocene.

> The fossils... constitute a curious and heterogenous collection of types... Such forms as the marmot, the camel, the antelope and the ostrich seem rather out of their due place. Possibly they were accidental wanderers along the plain, unless we admit that the plain itself was the steppe, then elevated considerably above the present flood plain level.

Also:

> Though these pioneers probably arrived with a knowledge that crude stones could be used in a variety of useful ways, it would seem probable that the lithic industry of Choukoutien was largely if not wholly a slow autochronous development; that the latter in any case was indeed an extraordinarily slow one, is witnessed by the relatively insignificant advances made in technique over the many, many centuries during which the *Sinanthropus* community must have occupied the great cave of Choukoutien...

The climate was mild. "Curiously enough, however, a generally effective faunal barrier seems to have existed then just as now, between the Yangtze and Hoangho basins." Just as now! Why not now?

There are some Mousterian (Neanderthal) cultural affinities: "As a matter of fact most of the... quartz specimens would seem to be indistinguishable from the major part of the quartz artifacts which have been collected in some of the Mousterian caves in France." Then: "There also occur throughout the deposit vast numbers of burnt and fragmented bones." Further, much of the deposit is of ashy and burnt clay of different colors, possibly of a great many fires, but also possibly of wind and water transported ashes. Almost nothing but cranial parts of Sinanthropus was found in the deposits, despite the abundance of mammalian bones in the thousands of cubic meters of debris examined. Could the skulls alone have been buried in the pit (a possible Mousterian practice)? Or washed in from a nearby settlement? One can conclude that more direct evidence supports a short-time life of the cave than a long-term history.

Yet pressure is exerted on the curators of the site of 'Peking Man' to re-date it to carry it backwards in time from 200,000 years to over a million years, so as to match East African specimens of homo erectus, which in turn has been found in association with australopithecus, and this extends backwards by another two million years, all based upon the validity of potassium-argon radiodating which is suspect. It is not beyond reason that this whole dating scheme will soon collapse and the hominids will be carried forward in time, leap-frogging the geochronological conventions of the 1920's, to the very edge of the holocene, a dozen thousand years ago.

FOOTPRINTS

At a site, G. Laetoli, Tanzania, the fossil imprints of three individuals, thought to be gracile australopithecines, were discovered in a consolidated tuff of volcanic ash dated by the K-A method at 3.6 to 3.75 million years. A stereometric camera was used to compare the footprints of these two individuals with modern footprints. The contour patterns are similar. The impression of the heel, ball, arch and big toe are similar. "The pattern of weight and force transference through the foot... also seem to be very similar."[60]

A lucid description of the K-A dating technique is to be found in *Lucy* (187-207). Johanson and his collaborators worked hard on Lucy's K-A dating of three million years to reduce the "margin of error" from 200,000 to 50,000 years. Then, on the basis of new information coming from paleomagnetic matching of rocks here and elsewhere and matching of dated fossil pigs found in rock strata of the same type elsewhere (biostratigraphy), they discarded the 3m/ y date for a new older date of 3.75 m/ y. Lucy became 750,000 years older.

One can scarcely be surprised if the reader, at first awfully impressed by radiochronometric machines, becomes now disenchanted when these are abandoned for divination from pig bones. Perhaps Lucy is a million or two years on the younger side and was gassed with her friends in a recent volcanic oven. And maybe the footprints at Laetoli were made by the Leakey family on an outing, before they had their first foundation grants. But this we know cannot be, for Mrs. Leakey would remember whether the volcano was then active.

---

[60] M. H. Day
and E. H. Wickens, "Laetoli-Pliocene Hominid footprints and bipedalism," 286 *Nature* (24 July 1980) 386-7; R. L. Hay and Mary D. Leakey, "The Fossil Footprints of Laetoli," *Sci. Am.*, Jan. 1982, 50-7; and on Lucy's new age.

The age of Lucy did not long stand where Johanson had placed it. In 1982 Boaz and others made new faunal comparisons that younged her and her earlier Afar associates by half a million years, and F. H. Brown compared volcanic tufts and likewise found Lucy much younger than she had seemed to be; a basalt testing at 3.6 m/ y lay above a tuft of 3.2 m/ y, the basalt test, less reliable, was superseded.[61]

Johanson could recognize his shoeprints and a cigarette package in the wadi where he had worked two years before, for there had been no rain. Yet "we surveyed the 333 site. A good deal of sandstone had crumbled down from the overburden above. It was now scattered in large blocks and smaller chunks over the hillside that had been so carefully screened for fossils two years before." Two years and 3.75 million years: close to two million times that amount of debris might have been dumped in the area since Lucy's days, even with a uniform climate (which he claims) and no natural disasters to muck it up (but 10 volcanos were active thereabouts in Lucy's days).

Old or young, the hominid and *homo* types have overlapped in time and habitat, as well as in numerous traits. Michael H. Day writes:

> It has been pointed out by a number of workers that the approximately contemporaneous Ternifine mandibles (jawbones) of Algeria and the Peking mandibles of China show extreme similarities; the great similarities between the Peking femurs (thighbones) and the Olduvai Hominid 28 femur have also been noted. A reasonable explanation of this similarity is that migratory hunting patterns had brought many groups of Homo erectus into contact and that exogamous (marrying outside the tribal group) breeding patterns had resulted in the widespread occurrence of certain traits. These similarities are very likely too great and consistent to have resulted from separate evolution along parallel lines in isolation; and, indeed, the degree of similarity seen in the available material makes it extremely unlikely that long-term isolation was a factor in human evolution after the early middle Pleistocene.[62]

Ashley Montagu long ago pointed out that Swanscombe man, who was quite modern, preceded Neanderthal, and that a Swanscombe type was found at Quinzano, Italy and placed in the Middle Paleolithic. Also before Neanderthal came Fontechevade man, with cultural remains, and he "would appear in all respects a modern type of man."[63] He alludes to Louis Leakey's

---

[61] See Boaz in 300 *Nature* (1982) 633, Brown, *ibid.*, 631.

[62]*Guide to Fossil Man,* 1956.

[63] "Time, Morphology, and Neoteny in the Evolution of Man," 57 *Amer. Anthrop.* (1955), 15ff.

Kanam and Kanjara discoveries as modern but Middle or Lower Pleistocene.

AMEGHINO'S ARGENTINE HOMINIDS

The extensive works of Fiorentino Ameghino, the Argentine paleontologist and archaeologist, are due a review in the light of recent oceanography, paleontology, and anthropology. During his lifetime he was attacked and ridiculed; he lost his university position for his ideas; nor has his fame been restored to this day. Several of his claims, apart from the many new species of extinct animals that are accredited to him, are beginning to ring true.

He proposed, on the basis of numerous explorations and excavations, that man had existed, with an Acheulian culture, in the Pliocene period and earlier, an age that only now is being invaded by East African hominidal discoveries. He found human remains, tools, and habitats associated with the giant fauna that were extirpated at the end of the Pleistocene. He found carapaces of giant turtles, with diameters around 1.5 meters, that could house dwellers of the plain, and inside of them, flint tools and selected bones; man, he thought, used these carapace homes on the treeless plains to avoid the giant animals of the age. He could not but believe that the association of man and great animals stretched far back into the Pliocene, even into the Miocene, and possibly the Eocene.

He argued vehemently for the existence until recently of land bridges between South America and Africa, actually in the time of man. No doubt that he would have welcomed the theory of continental drift in vogue today, although he followed a theory with other well-known writers, that the land between the continents had sunk, rather than split up and drifted.

His most shocking hypothesis was that mankind had originated in the pampas of southern South America and had moved North and East across continental connections. He called the Central Atlantic bridge the "Guyana-Senegal" connection. This is also the Antilles-Mediterranean link, which Suess, Lapparent, and other geologists and paleontologists perceived to exist in the Tertiary period.[64]

"I believe," he wrote, "that one can regard as susceptible to nearly rigorous proof the following facts: 1. The American population is not a

---

[64] "L'homme préhistorique dans la Plata," and "L'âge des formations sédimentaires tertiares de l' Argentine en relation avec l' antiquité de l'homme," in *Obras Completas* 24 vols., (Buenos Aires 1912-36), vol 2.

unique and homogenous race but the product of crossings of different races. 2. One finds individuals and tribes representing races of the Old World, but the mass of people is distinctly different... 5. Emigrations from the Old World always found the Americas peopled by natives... 7. While Europe was still peopled with savages, America possessed very advanced peoples living in great cities and constructing grandiose monuments. 8. At different periods, new emigrations took place toward the Old World... 10. The most ancient peoples of Europe, Africa and America were in communication. 11. The communications were facilitated by land, today disappeared. 12. The existence of this land can be demonstrated by tradition, prehistory, archaeology, ethnology, linguistics, philology, anthropology, botany, zoology, paleontology, and geology. 13. Up to now, science has not been able to determine in what corner of the globe man or his precursor made his appearance for the first time."

Ameghino describes skeletal material and crania from the Canyon of Moro (North of Necochea)[65] as of a people rather over four feet tall, long-headed, prognathic, small-brained, small-toothed, and generally exhibiting bone-structures foreign to modern man. He called this group of "hominids" *Homo sinemento*.

In another paper, Ameghino and his brother describe an apparently incised Protorotherium jawbone that they discovered. This would place Patagonian man over thirty million years ago, in the Eocene age, far earlier than the most radical of present-day datings which range up to five million years, and then only hypothetically. Two famous anthropologists from the United States visited the site, Ales Hrdlicka and Bailey Willis; neither accepted Ameghino's early datings of man or even the presence of a hominid in the Western Hemisphere, much less the four races of hominid that Ameghino claimed to have discovered.

Since the present author has not studied the problem extensively or at first hand, and indeed the materials for such a study may no longer exist for specialists to investigate, one can only remain in a state of mystification, hoping that the search for primordial humans in South America will be vigorously pursued.

---

[65] "Descubrimento de dos esqueletos humanos fosiles en la Pompeano inferior del Moro," *op. cit.*

METHODOLOGICAL POSSIBILITIES

Oxnard's statistical, computer-assisted techniques of comparative anatomy might well be applied to test new hypotheses. They are especially adapted for logical operations in which time should be squeezed out. Pearl computed coefficients of variations in the human species, along seventy dimensions. G. Simpson deemed the results to show a not unusual variability in comparison with other mammal species. [66] The data, he thought, indicated that man was changing rapidly. If modern man is so variable in physical structure, it can be assumed that fossil men (hominids included) will also be at least as internally deviant, and in fact they are, even if the australopithecines and homo erectus are examined separately.

But now let us group Neanderthals and proto-modern types with modern man, and australopithecus with homo erectus. The number of parameters of difference within the two groupings will probably remain the same - the aforesaid seventy perhaps. The variations or values within each grouping will increase. What are the two sets of coefficients of variations? What are their means and extremes? Are all of these indices equal within the two groups?

Then plot the ancient aggregate against the modern aggregate on every parameter, and on the means and extremes. Calculate all the differences and principal sets of differences and express them statistically. Test then the following hypotheses: a) The internal differences of the ancient group are of the same mean and range values as those of the modern group. b) The differences between all individual values and sets of values of the ancient and modern groups are not significantly greater than the internal differences found in each of the two groups.

Both hypotheses are deemed to be supported if the differences trend toward their confirmation. If the hypotheses are largely confirmed, elapsed time between ancient and modern man must be presumed to approach zero time.

The conclusions thus derived are subject to attack from 1) Independent measures of time by geochronology and any evidence of an independent archaeological kind such as aberrational cultural developments, as well as by 2) Independent knowledge from evolutionary genetics, from evolution by other means such as natural selection, and from paleontology concerning the length of time that the traits under examination require to reach their extreme parameters. If neither kind of independent control is valid and reliable, beyond the limits to which the aforesaid tests of the hypotheses are

---

[66] *The Major Features of Evolution*, N. Y.: Columbia U. Press, 1953, 78.

valid or reliable, then the hypotheses may be maintained: The groupings of ancient and modern man are internally homogeneous; elapsed time between ancient and modern man must be very short. Since little of this proposed work has been performed, however, the value of the hypotheses must be temporarily judged on the basis of such logic and evidence as are otherwise presented in this chapter and book.

TIME UNNEEDED FOR CULTURE

Oxnard is impressed by the uses to which a long history of mankind might be put:

> knowing as we do the enormously greater speed of psycho-social evolution as compared with the slow rate of biological evolution, then a larger absolute time span of, say, 5,000,000 years, may allow an even greater amount of relative evolutionary time for the evolution of the behavioral, cultural and intellectual qualities that stamp man as unique from any animal.[67]

But whoever said so much time was needed for cultural evolution? We shall soon be arguing that culture was practically instantaneous.

Some old evolutionists gave 50,000 years as the age of modern man. They were thinking in physical, not cultural, terms. That splendid hoax, Piltdown man, was expertly placed at 500,000 years and then a few years later just as expertly placed 50,000; finally, of course, he achieved the surreal, a timeless mockery of scientoid pretense. By the newest estimates, mankind would have had one hundred times as long as these 50,000 years to rise from some non-human level to its present state.

To insist that very old fossils of modern physical type must have had a culture provides a sword that cuts both ways against time. The physical as well as the mental traits of the homo species, if deemed to imply each other, might be dated very recently. Homo sapiens might be born within hailing distance of 14,000 B. P., a basepoint that I have developed in *Chaos and Creation* for the Holocene age.

To allow quantavolution in a short time, one must agree that some part of evolution might be systemic, that is, permit a set of crucial human changes to occur together in the same moment and perhaps by the same instant mutation. The issue has been hotly argued. A plurality of biologist

---

[67] *Op. cit.*, p. 122.

are point-by-point evolutionists; very few are saltationists, quantavolutionists or systemists; many are puzzled over the great variety of points to be covered over time, no matter how long, and yet unready to accept "successful monsters" as the answer.

There is no way of soothing the bafflement and frustration concerning measures of time. I have mentioned traditional geochronology and potassium-argon radiochronometry as the bulwarks of long time reckoning. Probably I must say more of them here inasmuch as they are accepted with little question by some of the foremost paleoanthropologists.

Traditional geochronology needs to be considered mainly because it offers a fall-back position, should radiochronometry be deemed invalid. The major drawback of geochronology in regard to fossil man is that time is measured by evolution; the time scale follows the fossil record of the sequence from "lower" to "higher forms."

The defensive positions of a century ago are irreparably in disrepair, however. At that time the age of the Earth itself was being argued in the highest scientific circles in the neighborhood of thirty to ninety million years, which would on today's hominid reckoning give perhaps one-tenth of all earth-time for the development of man.[68] But then man was still hovering in the five figure bracket of 20,000 to 90,000 years. Certainly, were it not for radioactive dating methods, evolutionary theory would be at an impasse for lack of time for mutation and for natural selection to transform the biosphere.

Like question-begging is the plague of natural selection, circular reasoning is the plague of traditional geochronology. "The rocks do date the fossils, but the fossils date the rocks more accurately... circularity is inherent in the derivation of time scales."[69]

There are neither transition fossils in any number to mark the important fossil stages, nor complete fossil columns showing the evolutionary sequence; nor is evolution a hard set of facts. Yet index fossils with a doctrinaire chronology are imposed on the rocks and the rocks assigned dates. Then rocks of comparable type, though lacking fossils, are dated accordingly, and many of the strata and formations surrounding them, too.

---

[68] E. G. J. Joly measured the runoff of sodium into the oceans to get "An Estimate of the Geological Age of the Earth," of 89 million years. Smithsonian Institution, Annual Report, 1898-9, 247-88. Melvin Cook, *Prehistory and Earth Models*, London: Parrish, 1966 criticizes a number of such techniques. Also A. de Grazia, *Chaos and Creation*, Princeton: Metron Publications, 1981, ch. III.

[69] J. E. O'Rourke, "Pragmatism versus Materialism in Stratigraphy," 276 *Am. J. Sci.* (Jan. 1976), 51; H. M. Morris, "Circular Reasoning in Evolutionary Geology," Institute for Creation Research, no 48, June 1977, iv.

Velikovsky has ingeniously displayed, using Blanckenhorn's study of the Syrian-Palestinian rift valley, through which pass the Jordan River and Dead Sea, that the old geochronology, before radiochronometry, could properly formulate for it a history of a few thousand years, rather than many millions of years. [70] He further used proto-historical evidence, that of Biblical sources, to strengthen the theory of short duration for the rifting of the area. The older methods of geochronology are often too flexible to engender confidence.

We must bring time into a new order. So long as it is the tool of the old vision of a point-by-point development of humanity, time will stretch out of bounds. The Holocene-Pleistocene boundary is not fixed upon an event, unless it be an end of the ice ages. But the ice ages are still going on, and it is doubtful that they played much of a role in the humanization and diffusion of man, except for imposing sometimes rather obvious limits upon settlement. The Pleistocene-Pliocene boundary was set by the International Geological Congress of 1950 on the basis of late Cenozoic stratigraphy in Italy, more precisely on the entrance of northern marine invertebrates into the Mediterranean. This boundary, too, is scarcely useful, and should be ignored in reckoning the origins of man in time. The Pleistocene record is always discontinuous and fragmentary, especially in glaciated areas. The task of scholars "would have been incomparably easier if some stratigraphic section covering the entire Pleistocene were available, showing, for instance a complete sequence of alternating tills and soils. Unfortunately, such a section seems to be available nowhere in the glaciated areas."[71]

We note, too, how geological time-reckoning expands as we go back in history. The Upper Paleolithic artistic period was dated back 30,000 years by French scholars and geologists, working on remains in caves and rock shelters. Estimates of sedimentation rates of deposits into which artifacts were sandwiched, gave such duration. But the dating of the Upper Paleolithic artists is more a working consensus that an absolutely tested fix. Pergrony and Caslis give us an age of 4500 years ago for metals, a Neolithic lasting 5000 years before then, a Mesolithic of 2500 years, an Upper Paleolithic of 30,000 years, a Middle Paleolithic of 80,000 years and Lower Paleolithic of from 800,000 to 1,500,000 years.[72] As we have pointed out, this last figure is now verging upon five million years.

The Upper Paleolithic period falls between the claimed periods of competence of radiocarbon dating and potassium-argon dating. The most

---

[70] "The Destruction of Sodom and Gomorrah," VI *Kronos* 4, 1981.

[71] C. Emiliani, "Dating Human Evolution," in ii *ED.* 59.

[72] *Notions de Préhistoire,* Perigeaux, 1975, 11.

careful work on this period is therefore dependent on sedimentary dating in large part, and this cannot get around the possibilities of periods of flood and torrents, laying down blanket after blanket of clay and gravel to create illusions, in today's peaceful landscape, of the passage of much time. This is no new problem. For instance, when Alfred Wallace was writing his studies of the distribution of animal life in the nineteenth century, he had to confess to the great difficulty of judging sedimentary deposits.[73] In repeated discussions at the Dordogne cave and shelter sites with French scientists who have excavated and are responsible for them, I have been unable to accept their meticulous reconstructions as valid.

In the end, they rely nowadays upon carbondating, which although it often upsets their expectations, at least keeps them in the Paleolithic period rather than moving them into more recent times. That radiocarbondating which is based upon measuring a ratio involving the diminishing amount of carbon-14 isotopes discoverable in organic remains, can be erratic, owing to atmospheric, species, and soil transformations, has already been the subject of investigation. Recently, changes in the Earth's geomagnetic field have been added to the several conditions that alter radiocarbon dating. Unfortunately, the usefulness of radiocarbon dating decreases exponentially as we move into the periods of the Neolithic and beyond, when the need for a dating instrument becomes increasingly acute.[74]

Geologists bought evolutionary time to preserve themselves from alternative catastrophic hypotheses. Whereupon the biologists and anthropologists, together with the geologists, were persuaded of radiochronometry by geo-physicists. The Potassium-Argon test claims validity over a time span of a billion years and more, beginning at 100,000 years or less before the present. Its favorite rock for testing is erupted volcanic material, ashes and lava. It establishes a constant rate of decay of the isotope potassium-40 into the isotope argon-40 (40K to 40A). Then it measures the amount of 40K and 40A in a rock sample and, by the proportion of the two, determines the 'age' of the rock, hence of fossils embedded in the rock. A high proportion of Argon-40 signifies an old age.

Unfortunately for its validity, and despite the brilliant technical theory and achievements represented in its applications, the 40K ug 40A test suffers from a defect common to radioactive elements in nature. The

---

[73] *Geographical Distribution of Animals,* N. Y.: Harper, 1876, I, ch 1.

[74] M. Barbetti and K. Flude, "Geomagnetic Variation during the late Pleistocene period and changes in the radiocarbon time scale," 279 *Nature* (17 May 1979), 202-5. See *Chaos and Creation,* ch. 3.

elements migrate. In consequence, the proportions change, giving illusory ages. Rocks can both acquire and lose both elements, or either alone.

Moreover, one cannot rely upon a temporal sequence that appears nicely to show older strata succeeded by younger strata as a proof that the sequence occurred smoothly and without disturbance. For the whole sequence may have been laid down in short order during a turbulent period that is accompanied by high argon deposition, or the eruptive sequence of a volcanic source can lay down deposits, first heavier, then lighter, in Argon-40, owing to a tendency of such trace materials to migrate from heavier to lighter rock. It may not be necessary to disbelieve absolutely in the validity of 40K ug 40A dating to maintain a quantavolutionary opinion of the process of humanization. However, it is more difficult to explain certain critical fossil data and the mechanics of humanization while adhering to a long time perspective. Vast stretches of non-eventful time have to be accepted between the occasions of significant changes, such as bipedalism, large brain, tools, and language; or else the finest, minutes, multitudinous ladder rungs or steps are forced upon one, leaving one again in baffling contradictions and a need to search for a meaning behind evolution such that every bit of change requires every subsequent bit of change, connecting intelligence with depilation, and so, on, thus accounting for the confusion of ladder-rung-labelling, with now one trait, then another being given priority.

OLDUVAI GORGE

Homo erectus bones and artifacts, which may even be australopithecine, have lately been discovered in the Syrian-Palestinian rift valley that we have already claimed to be of recent origin. In a letter of October 15, 1981, Professor Ernst Wreschner of the Department of Anthropology, University of Haifa, wrote me that at Ubeidiya, "together with an industry of pebble tools, spheroids and primitive handaxes they found a skull fragment and a tooth. Not enough to say *Australopithecus* or *Homo erectus*. I tend towards the latter. Supposed time: *ca* 800,000 years. Because of the similarity with Olduvai 3 it became designated: Israel-Olduvai. The Ubeidiya site, the time of its occupation *ca* 800,000 years ago and till about 250,000 ago, was a lakeside camp, before the tectonic tilting. The living floor is now tilted *ca* 43 degrees." (Dating was by 40K-40A of underlying and intermediate basalt (lava) layers, thus similar to E. African practice generally.)

But, now this tool-strewn "Ubeidiya" hominid site of Israel has been reevaluated with respect to homo erectus in Africa and moved from 700,000 years ago to 2 million years or more, placing it alongside or possibly older than any early Acheulian finds of Africa.[75]

> Here we evaluate fossil mammals from Ubeidiya, which are stratigraphically and directly associated with Early Acheulian artefacts, and find no substantial reason for considering the locality younger than 2 Myr, and possibly as much as 500,000 yr older than any record of Early Acheulian artefacts or Homo erectus in Africa.

In this book, I am suggesting that the Rift finds generally should be deemed contemporaneous, so that the new placement is welcome in one sense. However I also suggest reconsidering both homo erectus and australopithecus as quite young, that is, moving the Acheulian to the beginning of the Holocene period. In other books (The *Lately Tortured Earth* and *Chaos and Creation*) I ask, too, that geological dating methods be revised so as to allow the drastic younging of the strata in which all hominids and homo erectus are found.

These discoveries bear ominously upon the famous centerpiece of current paleo-anthropology, the Olduvai Gorge. The narrow floor and steep sides of Olduvai Gorge in Tanzania are a typical element of the East fork of the Great African Rift, which cuts from at least South-eastern Africa to the Red Sea. The problem of Olduvai man and culture is part of a complex world wide geological history that I have outlined in *Chaos and Creation*. I appreciate that I cannot here reproduce these materials, nor bring to bear more extensive materials analyzing the particular setting and criticizing the methods of radiochronometry employed.

I can only state the nature of the problem and alert the reader to the ultimate surprises that may be awaiting historical anthropology in this setting. To do so, I quote here from an exchange of letters with Dr. Melvin A. Cook, a geophysicist, recipient of a special Nobel Prize for his studies of explosives, and author of *Prehistory and Earth Models*. (1966). On March 10, 1976, I wrote Dr. Cook the following:

> ... Presently I am perusing the three volumes of reports by the Leakeys and others on the Olduvai Gorge. Here I think is a main intellectual battle front, and one that calls especially for your attention. Here the conventional paleontologists, geologists, radiochronometricians, and evolutionists are lined up in force. All the discoveries are squarely upon the Great African Rift, the bottom of the deposits is an igneous basalt, and from then on up for 300 feet are layer upon layer of tuff, clay, marl, Bonneville-like type 'sediments' and scanty soilroot elements, with earliest

---

[75] C. A. Repenning and O. Fejfar, 299 *Nature* (1982), 344.

fragments of *australopithecus* and *pithecanthropus* interlarded at 'living' sites, and with abundant mammalian and lake fauna including very large and modern species both extant and extinct. Hominid and faunal transitions are indistinct from bottom to top, similarly the abundant scattered artifacts. The discoveries are eroding off the walls of the rift and are also found by digging back from the walls. The whole is dated after some controversy from 2 million years at the bottom to about 300 thousand at the top, using K40-A40. Everyone is proud of this showcase of many disciplines.

On the other hand, I cannot but perceive a quite different solution, that is, the initiation of a heavy cone and fissure volcanism, the uplifting of the great plateau, a watered depression, successive floodings and lava flows and fallouts of ash and dense material from the many nearby centers of volcanism, repeated incursions of hominid and faunal species, and finally the rifting as a forking from the world global fracture. Several cultures, from Asia to Kenya, 'remember' the upheaval of the rift, Hebrews, Arabs, and Blacks...

If you would look at the excavation profiles in Vol. III, you will note an average of about 30 levels, most or all of which are probably turbulence deposits.

In his reply, dated May 5, 1976, Dr. Cook said, *inter alia:*

Interestingly enough, just a day before your letter of April 29 arrived, my son Krehl returned from Kenya and a visit to the Olduvai Gorge and Great African Rift valleys with a large group of geographers and geologists. He gave me his vivid first-hand impressions of the geology of this region and the occurrences of fossils. It coincided very well indeed with your descriptions. After his extensive study of the region of the Olduvai Gorge and the surrounding area, he said he is completely convinced that it should be understood in terms of a catastrophic continental drift with great global overthrusts and subsequent catastrophic readjustments that have really been the facts that have shaped the region... The global extent of the great rifts, their obvious relationship one to another, the sort of chaotic geology found in and around the rifts throughout the world - not merely those in Kenya, and the excessive fragmentation of fossil skulls and bones (human and animal) in these regions are the sorts of information that to us prove that the great rifts were created all at once, i e., catastrophically...

Your reconstruction of the situation in the Great African Rift and Olduvai Gorge is very plausible...

One *must* handle K-A dating, consistent with all the facts dealing with it, by simply dismissing it as unscientific and completely unreliable, indeed absurd. They simply don't publish the sort of facts they know about that would kill K-A dating once and for all if they are known. For example, I have heard that year-old volcanism in Hawaii can yield K-A 'ages' of several million years...

As I pointed out in PEM *[Prehistory and Earth Models]*, the A 40 found in igneous rock is largely *nonradiogenic* contamination. Leakage of rare gases from the crust is too great to permit any reliable dating. Moreover, leakage can both deplete and enrich. For example, leakage of A^{40} along such vast outscrapings as occur in the Great African Rift can concentrate A^{40} inventory of the earth even if the earth were five billion years old...

Should the hypothesis of the recency of Olduvai history become adopted, the theory of homo schizo would be strengthened. Should it not be acceptable, mysteries, contradictions, anomalies and confusion would persist, such as the astonishing million-year retardation of human implement development that I stress in these pages and that Sonia Cole, among others, refers to.[76] In such a case, the theory of homo schizo would need to retreat to a position asserting that the true human was born recently out of catastrophic events which allowed a further climactic mutation and/ or chemico-physiological transformation. We would have to abandon australopithecus and homo erectus throughout the Old World, with all of their humanlike traits, to live out very long existences sub-humanly.

A SURPRISING COLLAPSE OF TIME

But nothing stands in the way of objectively and empirically explaining the whole set of fossil hominids that rift excavations extending from Syria to Southeast Africa have produced as a short-term occurrence under catastrophic conditions. The same is true of Peking Man (see Index) and of all other hominid and protohuman finds, except perhaps certain 'anomalies' (to borrow the excuse of the opposition). The Olduvai Gorge hominids and homo can be readily brought into the Holocene period.

Consider how rapidly man changes, physiologically and culturally, under present-day observation and from our earliest direct knowledge, which is Upper Paleolithic and Neolithic. To thereupon add five million years (or 100,000 'memorial' fifty-year generations) of mental and cultural evolution to a substantially completed anatomical structure would reduce to absurdity the uniformitarian theory of the evolution of modern mankind. Or else man would have evolved, and been destroyed, time and time again, never being

---

[76] *The Prehistory of East Africa*, London: Weidenfeld and Nicolson, 1964, 130.

extinguished. (But of course this would be another form of universal catastrophic theory).

So much time is not needed, if man is evolving on a consistent anatomical base. More time is now defeating to evolutionary theory; the evolutionists do not yet appreciate that they have crawled out farther and farther on a limb which may suddenly and soon break off at the trunk. For instance, could humans and hominids have lived for millions of years without having reached the Americas, where elephants, camels, horses and other mammals abounded? Would they have waited until 100,000 years ago to descend upon Australia?

Nor can evolutionists cease to stretch time and beat retreat to shortened time. If the time is drastically shortened for paleoan-thropology, the radio-dating techniques collapse. Then all which depends upon the techniques -- prehistory, paleontology, geophysics, geology, climatology, etc. -- will come under revolutionary assault.

Against the background of this stupendous reversal in prospect, other conclusions about fossil man pale. The australopithecines existed alongside homo erectus and other types of man, as well as many kinds of ape. The vanished hominids were destroyed by or adapted to a dominant strain of the human race, in conjunction with natural catastrophes. We shall consistently maintain that homo sapiens schizotypus (catastrophized *homo sapiens*) reduced his live, physiologically compatible brethren, whether australopithecus, or homo erectus, or homo sapiens, to subjection, or he exterminated them.

CHARDIN'S ORTHOGENETICS

By now it should be clear that we are heading implacably toward a theory of biological quantavolution, an eventful scene in natural history, where a hominid walked upon the stage and a human walked off. This hovers upon creationism, in the theological sense. But it is not such. Nor do we need to employ here orthogenesis such as Teilhard de Chardin calls upon, a divine or even natural penchant of the soma of a species to transmute into a phylum crowned by a mysterious *noos*.[77] It is not that we want to, or can, or must take away from humankind all the glories that we claim for it. But this matter is not germane, and there has always been an abundance of literature exclaiming upon the incomparable and marvelous capabilities of homo

---

[77] *The Appearance of Man*, N. Y. Harper and Row, 1956; *The Future of Man, ibid*, 1964.

sapiens sapiens. We say that humanization is a brief episode, accomplished by a set of minor alterations, and followed by a mighty effect.

De Chardin was close to such significant events of fossil anthropology as the fraud of Piltdown Man and the excavation of the caves of Choukoutien in China that gave up the skulls of Peking man (sinanthropus); he was a Jesuit and a social philosopher, playing a role rather like that of Loren Eisely in America. He accepted uniformitarianism but yet conceived of teleology in evolution. He thought that the Peking skulls that were found throughout the whole fifty meters' depth of a filled fissure of breccia, ashes, and clay, along with many extinct animals, were of thinking humans. He saw evidences, as did others, of fire-making and deliberately chipped stones. The time of occupation was estimated, by himself and others, at between 100,000 and over one million years. Two fatal observations, that are conveniently evaded in most discussions of Peking man these days, are that the assembled material may have been catastrophically collected and impacted in a short time and that the skulls may have originated elsewhere.

Peking man, later identified with a widespread group of hominids of the homo erectus designation, was part of "the trajectory of a humanity moving persistently towards ever higher states of individual and affective consciousness."[78] However, it seemed to him that this hominid group died out in the Middle Pleistocene, then estimated at some 200,000 years ago, as did the more primitive but possibly also pebble-chipping australopithecines, which have also been found over half the Old World.

De Chardin found himself trapped between microevolution, point-by-point changes, which he nevertheless calls quantum jumps at one place[79] no matter how small they may be, and macroevolution, a large quantum leap. But he could not imagine the form of the leap except that it might be "a simple chromosomatic mutation" and that the gap between the human and the australopithecine "has not necessarily been greater, in size, than that ordinarily observed or stimulated, beneath our eyes, in animal or vegetable populations at present living. In the case of man, we seem to have an example of mega-evolution governed by chromosomatic play of a perfectly normal type." Yet the germplasm is orthogenetically prepared for "the great leap of hominization" and cerebration.

This is *ex post facto* reasoning of a dubious kind, made necessary because Chardin feels he must have a marvelous (teleological) cause. Ultimately he would then argue for *noos* or spiritual intelligence, and soul, detouring

---

[78] *Appearance*, 123.

[79] *Ibid.*, 136-7.

around all that is known about the brainwork and central nervous system, not to mention the behavior of humankind.

DOBZHANSKY, SIMPSON AND QUANTUM EVOLUTION

Theodosius Dobzhansky picks up the problem of the quick leap, too, in two sentences of his masterful treatise on *Mankind Evolving.* But he perceives the leap as involving many quick successive changes. "Quantum evolution, emergence of novel adaptative design, may involve breaks in the evolutionary continuity when the differences between the ancestors and the descendents increase so rapidly that they are perceived as differences in kind."[80] He passes on to other matters, missing the chance, as does Teilhard de Chardin, of launching into a quite new paradigm.

After discussing, as if they were successive, a set of evolutionary are at least behavioral changes in prehominids, he raises the question

> as to whether upright stance, tools, monogamous family, change in food habits, or relaxation [sic] of male aggressiveness came first. Obviously we cannot answer with certainty, but it is most likely that these changes went together, with mutual reinforcement. What we are dealing with is the emergence of a whole new evolutionary pattern, a transition to a novel way of life which is human rather than animal. This is an example of an infrequent type of evolutionary change, which Simpson (1944, 1953) has called 'quantum evolution.' Evolutionary alternatives in general, and especially those in quantum evolution, are unlikely to involve changes of one trait at a time. The whole genotype and the whole phenotype are reconstructed to reach a new adaptative balance.[81]

This passage is remarkable in that, whereas Dobzhansky's work as a whole epitomizes the conventional uniformitarian and long-term evolutionary approach to the origins of human nature, here he is practically giving away the show to quantavolution. Any being that can perform all of these operations can and must perform all other human operations; man is born.

At one place he says that pre-man separated from apes "no less than 11 million years" ago.[82] He places the proto-homo sapiens at perhaps a quarter

---

[80] *Op. cit.,* 213.

[81] *Ibid.,* 209-10.

[82] *Ibid.,* 183.

of a million years ago. Presumably long periods of evolutional impetus occurred, or thousands of other changes took place before the sudden transformation. Then why is the great leap needed?

So it was Simpson who had originally muddied the still waters of uniformitarianism. What had he said? Reluctantly, with a step backward for every step forward, Simpson applied the term "quantum evolution" to the relatively rapid shift of a biotic population in disequilibrium to an equilibrium distinctly unlike an ancestral condition.[83] However, "the genetic processes involved do not permit making the step with a single leap." Agreeing with earlier work of Dobzhansky,[84] "the accumulation of small mutations is not only adequate to permit rapid evolution, such as involved in quantum evolution, but also the best substantiated mechanism for this."

The "small mutations" of a rapid type he accounts for by the availability of unoccupied ecological niches and the break-up of sub-groupings of a species into isolated pockets, so that one, which is preadapted, can change swiftly to exploit the niche, while the other groups often die out. Thus, some horses grow big, strong teeth while browsing, without needing them, but then, before the big toothed horses, isolated and with browsing overdone, could become extinct, the new form begins to use the teeth to graze rough grasses, then expands to fill the new niche. Simpson grants that his examples were not on large changes that bring in families, sub-orders, and orders, but thought that a process like this could cause the large changes.

Indeed it was because of the continuous puzzle of large-scale extinction followed by fully developed new species, mega-evolution, that he felt the need for a new concept. The hot breath of quantavolution was on his neck, but never could be let himself turn and face the concept. Nor could Dobzhansky or Chardin. They resorted to equivocation, denial and evasion.

When, later on, reports accumulated, that characterized the boundary-periods between extinctions and new species as times of natural catastrophes, Simpson resisted attempts to take up and enlarge his idea. He says in one place, "Since the groups involved in the major, more or less revolutionary episodes are highly varied in structure, physiology, and ecology, it seems unlikely that the intensified factors are the same for all of them."[85]

And then, farther along, he writes, "The real point is simply that a modified, relatively mild and gradualistic form of revolutionism is in accord with our present knowledge of biohistory, but that neocatastrophism is

---

[83] G. G. Simpson, *Tempo and Mode in Evolution,* N. Y.: Columbia U. Press, 1944, 206ff.

[84] "Biological Adaptation," 55 *Sci. Mon.* (1972), 391-402.

[85] In Albritton, ed., *Essays in Evolution and Genetics,* 291.

not." He likes 'neorevolutionism'. Perhaps this was because the notion of catastrophe, when fully realized as in the theory of quantavolution, affects seriously the theories of evolution, natural selection, and long-time natural history.

We can allude to a final example, one from primate history, based on a chart which can be found in Buettner-Janusz's *Origins of Man*[86] . The families of primates have clearly boundaried histories, with little overlapping from one age to another. Six out of seven boundaries are sharply defined by extinctions. Of course, the families may have quantavoluted at these points, rather than extinguished. If so, so much the better for our theory.

Where the boundaries of the geological ages are not clear -- such being actually the case -- the primate families themselves delineate by their careers the period boundaries, without the help of other fossils and rock strata. We note also that most of the living taxa have fossil relatives who became apparently extinct (or did they hide themselves somewhere?) eons ago. Quantavolution is manifested throughout the ages, but perhaps the ages are not so far gone either, and quantavolutions have been frequent.

Under the circumstances, a close look at the mechanisms that might produce humanization is justified. Time, period boundaries, evolution, culture, geological strata and types of humanoids -- all have begun to whirl about in our minds and we begin to wonder when the skies, too, will begin to whirl, and wish that we might have a theory - even if quantavolutionary - to stabilize the scene.

---

[86] *Op. cit.*, Figs 7.1, 7.2, pp101-2.

CHAPTER III

MECHANICS OF HUMANIZATION

If time were collapsed into a short span, cultural traces now deemed hominidal would appear human. Since footprints and bones that are deviously connected with the artifacts are also now considered close to modern man's, we may suspect that homo erectus and australopithecus, if not human themselves, of having had some human cousin living up North, say. This homo schizo would send his relatives fleeing east and south from the common ancestral home, which might be at the junction of Europe and Africa. After consolidating his position -- mainly defining his proto-mind and proto-organization -- he would reach out to contact them.

Then he would extirpate or interbreed with them, and come to dominate the *homo* line and promptly diffuse to the geographical limits of the world. Only the docile and miscegenable would be spared. Even today, several strains of homo schizo are in danger of extermination -- the pygmies, negritos, numerous Amazonian tribes as well as other Amer-Indians, and the Eskimos, for example.

If homo erectus and australopithecus were human, which we deem likely, then we should look for a hominid (X) as our ancestor. This 'X' might be much like man or a surprisingly different type. Granted that the probability is low, because he still has to appear alongside the fossil australopithecines, even modern man - to all physical appearances - might be his own ancestor. But, too, 'X' might be an unseemly anthropoid.

Eugenics cannot say how great a change of type can occur under special conditions nor whether certain species are more capable of quantavolution than others. We note how often in the fossil record, some species change while others remain the same. And we have already been startled into the realization that the change from hominid to human may have been anatomically slight.

Since we have little evidence to suggest who 'X' might be, we might leave the search for him in more knowledgeable hands, and assign ourselves the task of determining theoretically how such a hominid could become human.

We have already ruled out most of the traits that scholars have joined to the ladder of evolution - skeletal, muscular, sensory, alimentary, sexual and lingual-pharyngal mutations - as the crux of humanization. We have ruled out as well the growth of the cranium. We should also rule out the piling up of reinforced primate experience in a growing storage-box brain that would eventually begin to expel human products.

One need only contrast the races of mankind to see how little difference so many changes do make in psychology and behavior. With skin color from black to pink, hairiness from hirsute to hairless bodies, height from the very tall Watusi to the neighboring Pygmy, nose from flat to hooked, head from broad to long, cranial capacity from 830 to 2000° cc., differences of dentition, of blood groups, and so on, homo schizo has nevertheless come to possess a similar array of psychological qualities whatever his outward appearances.

We should look most closely for signs of self-awareness, of a split ego, for from this, we believe, and only from this, would come the flood of fear, the insatiable demand for self-control, and the outward movement of this need to control, taking the form of showers of displacements that would be transformable into human conduct. Symbolism would be the necessary external manifestation of the inward symbolizing needed to tie together the ego that had been split asunder. We would expect our newly quantavoluted person to behave recognizably as an imaginary Hominid 'X', close to the chimpanzee, in that his basic needs would be the same. Much of his behavior, too, might appear instinctive.

What would become quickly a critical difference would be an unending stream of delayed and unrecognizable stimuli in great numbers. He could be interpreted as an animal trying in amazing ways to consummate a new kind of stimulus-response, where the responses were delayed, as much as he might try to speed them up. He would be an action-delayed, hence decision-craving creature.

Whatever its cause, the character of the mutation may have been quite simple, confounding high-flown speculations that have adorned debate about human nature over the centuries. It may have been what Dobzhansky called a polygene mutation, carried over into many chromosomes, providing a slight quantitative, not 'qualitative' change, but yet a change with great effects. A systemic delay of microseconds in overall signal transmission in the brain might act as a suppressant of instinctual response, set up an echo of the self, and excite perennial hyperendocrinalism. The gestalt of creation (treated in the next chapter) would promptly take effect.

Besides mutation, it is conceivable that an environmental constant may have changed, provoking a human response that must continue as long as the constant remains unchanged. Further the human mind may have quantavoluted culturally because of experiences so intense and memorable that a new kind of creature emerged from them. We must look into these possibilities more closely. But before we do so, I ought to stress the importance of natural catastrophes as a background and source of quantavolutions in biology.

ANCIENT CATASTROPHES

Legends everywhere carry stories of great numbers of people reduced to a few survivors. They are obsessive tales, repeated continuously over thousands of years. Psychiatrists will readily deem them to be founded upon strong memories, a possibility that most historical scholars are loath to admit. The English prehistorian Childe says that the population of England in Paleolithic times numbered only in the hundreds, during Neolithic times in some thousands. Were then all the people of England mustered to drag the Neolithic megaliths of Stonehenge into position, and then reassembled in Ireland and Brittany for the same tasks?

Mircea Eliade, whose research on myths worldwide is justly renowned, is most impressed by the obsession of all peoples with the earliest times of creation, by the permeation of the totality of their cultures with the same obsession that great and terrible events occurred; yet he has not ventured to say that anything at all happened then. Are the legends mere fantasies of primeval poets, primordial Dante's, whose plots no later poet could ever improve upon?

Dobzhansky, of whom we have spoken, takes one sentence in a large book on human evolution to dismiss obsessions of creation as a 'natural' reluctance of people to conceive of infinity. Are peoples, (using his own perspective) supposed to recall their lives as apes? He and many others arrogate to an illuminated modern mind the right to conjecture and endorse ideologically the concept that humans long were few and became many with extreme gradualness. For all the other people who have ever lived, and who claim by a kind of culturally transmitted history that their ancestors arose in large numbers and were wiped out in cycles of catastrophes and revival, no place is to be reserved in science.

The Holocene period itself embraces many more fundamental natural events than were once accredited to it, as the latest edition of the *Encyclopedia Britannica* points out. A start has been made in assigning quantavolutions to it, but the major allocations cannot occur until chronological methods are criticized and reformed. Even though their character as ages is not yet defined, the occurrences of catastrophes of continental, if not global, scope may have occurred in half a dozen sections of the Holocene period -- approximately 2700, 3400, 4400, 5700, 11500, and 14000 years ago. I have attempted to order and designate these epochs in the book, *Chaos and Creation,* Chapter Four. Intervening catastrophes of less global scope were common. Plato, Boulanger, Carli-Rubbi, Cuvier, Velikovsky, Schaeffer, Hapgood, Kugler, Schindewolf, M. Cook, Kondratov, Patten, Bass, Juergens, and other scholars of times past and present have exposed the revolutionary character of natural events in such ages.

For the moment, and so that argument may proceed along its main line, the extent of catastrophism of the past fourteen thousand years can be barely sketched. It would help conceptually to regard all expressions of natural forces of a destructive character witnessed by modern humanity as but the flattened tails of negatively exponential curves of catastrophism. These natural forces of the past worked through explosions of the legendary elements of "earth, air, fire and water."

Catastrophic action and effects were manifested over the whole globe. Their intensity was such that the potential destructive force of deviations in the motions of celestial bodies must be introduced into the equation. Legends (oral history, amnesiac fragments, and sublimated tales) assert abundantly the priority of heavenly forces as destroyers of the world on successive occasions. The locations of large meteoritic explosions are discovered in increasing numbers. Increasingly, geologists such as Ager slip from the grasp of earth-bound uniformitarianism, and astrophysicists such as John A. Simpson conceive of their realm as explosive and disorderly.

It can be asserted and defended that in the past fourteen thousand years, a disorder befell the solar system that terrorized and transmutated the sensible biosphere, changed the atmosphere, cleaved and ravaged the crust of the earth, altered drastically the sky and surface waters and destroyed or severely damaged every civilization up to the seventh century before the present era. During these fourteen thousand years, it can be argued, human groups spent one-third of their time in an environment of natural and social chaos and suffered intense physical and mental stress. Again I refer to the book, *Chaos and Creation.*

Continents were fractured; mountain ranges rose; crustal material was exploded into space; cataclysms of water, ashes, oils, gases, and fire rained from the skies; ice ages came quickly and avalanched, not melted, into oblivion. Oceans were created; seas were drained; floods raged in every direction up to the very mouths of highly placed caves; climate altered in a day and the atmosphere was deprived, enriched, and poisoned on numerous occasions.

No single mile of the surface of the world can be bored for its actual stratigraphic column without discovering it to be at some points a catastrophic column. No matter what part of the destruction can be assigned to the ages before man, some part of it has to be attributed to the ages of man. Settlements and civilizations everywhere, from the Arctic Sea to the Tropics, from Spitzbergen to Tiahuanacu, are now, upon exhumation, shown to have been the victims of such events.[87]

Beginning with the larger part of its surface that is below the oceans, the earth is a scene of global disaster, punctuated by habitable oases. Scientists have known so all along but, in good conscience, have refused obsessively to reveal the fact. Whether one observes the ash and debris of hundreds of ancient settlements which, as C. Schaeffer has said, are studiously ignored or whether one reports on the ashes of primeval human sites, where, comments H. T. Lewis critically, ash lenses in places like Shanidar are offhandedly treated as "ash middens" or hearths, the 'conspiracy of silence' governs the authorities. The Sahara, the Great Salt Lake (Bonneville) area, the Gobi desert, the arctic tundra, the sub-glacial Arctic and Antarctic regions, East Africa, Southwest Africa, and so on to many smaller locales are signs that what happened to Mars almost happened to Earth. Practically all species that became extinct, and whose careers might be followed by fossil evidence, became extinct suddenly, as for example, some three-quarters of all large animals at the end of the Pleistocene, fully within the time of man's cultural flowering.

It appears, therefore, that every hypothesis trying to explain the means of humanization must be developed within the historical bounds of natural catastrophe.

---

[87] This column is discussed in the author's *Chaos and Creation* and *The Lately Tortured Earth* and see I. Velikovsky, *Earth in Upheaval,* Doubleday, 1955; Harold T. Wilkins, *Mysteries of Ancient South America,* Secaucus, N. J., Univ. Press, 1956; Claude Schaeffer, *Stratigraphie Comparée,* London: Oxford U. Press, 1948.

THE HUMANIZING FACTOR

The closest that we can come to distinguishing a key factor in humanization is an instinct-delay system operating in the brain but serviced by the body's electrical and hormonal system. This could be called the humanizing factor. In baseball language, the animal in us has been forced to touch base several times before completing the circuit and scoring, and so the human ball game is on. A close investigation of instinct-delay (see *Homo Schizo II*) emerged with the theory that it is an effect of the specialization of the brain, with consequent transmission delays in coordination of the total brain and organism, and that there is an over-crowding of consciousness because of a spillover of repeatedly insistent messages taking alternative routes for delivery or ending up in functional *cul de sac*.

I shall try to formulate the process of instinct delay here in a manner that will assist in seeking the mechanism of humanization. Instinct-delay (D) is a function or ratio of the product of the mass of the brain tissue requiring service (M), and the specialization thereof (S), to the product of access facilities (the number of receptors or docks and the number of routes pursued by messages) (A), the input of electro-chemical signals (E), and the velocity of the work of transmission (V). That is, $D = f(MS/AEV)$. When all other variables are held constant:

1. If the mass of the brain increases, instinct-delay (D) will usually increase.
2. If specialization of the brain increases, D will usually increase.
3. If the number of message docks and the number of routes to them increase, D will usually decrease.
4. If the input of electro-chemical signals increases, D will usually decrease.
5. If the signals move faster, with less impedence, D will usually decrease.

However, the variables are not entirely independent, although we do not know the extent of their interdependence. Thus, an increase in brain tissue may not bring a proportional increase in docks and routes available to supply the tissue. Nor will the larger brain necessarily be supplied by an increased input of hormones, which come from several places, or electrical charges, to transport messages. Nor do we know whether the electric and chemical signals will carry on with their former speed or will move less rapidly, unless some other factor increases their speed (which may, for example, be an electromagnetic change in the state of the environment).

That is, we cannot identify precisely the agents, nor the cause of their behavior. All that we can feel confident of, at this point, is that there are here the rudiments of an explanation for instinct-delay, hence self-awareness, hence humanization. We believe, too, that such a system is capable of empirical verification and modification. Further, it seems to answer a need in science for a concept that will go along with most of what is known of human development and human nature, and will not lead us astray as we seek to understand how mutation and other mechanisms could have occurred. Finally the concept will pay a large profit when it correlates with the mental and cultural behavior of the human during and after humanization.

QUANTAVOLUTION VS. EVOLUTION

We return to the contention that mentation and culture have developed by small increments over millions of years. We find in it a subtle ideological attempt at cutting change into such fine bits that it will simply blow away and nothing will be left to explain. Let us address it nonetheless.

The human is compelled to behave humanly in both mind and culture. Once granted that self-awareness was a quantavolution rather than very slow evolution, mentation and culture must originate at once. So one should ask whether self-awareness came at once.

Self-awareness is a trait that varies quantitatively among humans; some people are apparently unselfconscious, until closely observed -- then time, space, gods, rituals, discipline, and anxiety appear, until it is obvious that their 'unawareness of self' is a catatonic suppression. In any human group, we invariably discover a capacity to be taught self-awareness among practically everyone. Even the hominids among us, if there are such, know how to 'put on a good act. '

But let us speak of the past. Could the self-awareness of the human species as a whole, regardless of whether homo schizo ultimately emerged, and granting that he did, be a matter of slow accretion, slightly self-aware four million years ago, more so three million years ago, even more so two million years ago, practically modern a half million years ago, modern 30,000 years ago? Whereupon, would not developments that require less self-awareness take less time, and inventions demanding more self-awareness take more time? The use of the right hand, for instance, might long precede the invention of tools for the right hand, and then another long period would be allowed for language and even this divided into words for sensible things, and later words for abstractions, then drawings, then domestication of animals, and so on.

All of this is not impossible. It is widely believed that some hundreds of physical and cultural changes were laid upon Hominid 'X' gradually over millions of years, and that the "flowering" of culture occurred among Upper Paleolithic man and then again in Neolithic times, and then again in the iron age, and once more in the recent centuries of science - these "flowerings" being expected as accumulative, branching effects.

However, the external environment and the internal tensions of homo schizo, the ones who were fully self-aware, would immediately have stressed the whole community to maximize self-awareness. The drive is socially contagious, and irresistible. It comes from the fear of itself and the need to control itself. It is not dealt out by a third party. It is excited by itself. Therefore it cannot emerge piecemeal. It must emerge for all it is worth as soon as it exists. A homo schizo in a group of Hominid 'X' would dominate or die.

But might the self-aware have been precisely those who gradually became such? No. Unless a guiding hand to physical evolution were present, we cannot expect this trait to have emerged in ever-increasing quantities by successive mentations, like a turning of the screw, each turn producing a higher level of self-awareness with a consequent output of new ideas, fabrication, and social inventions. Indeed, evolutionists teach us to avoid such pathetic notions. Who advocates such a guiding hand? The psychosomatic Lamarckians probably, and I may sympathize with them. But why should a beast will for himself a small increment of self-awareness, and then another and another, especially when the psychological effects of self-awareness are not at all comfortable, not even tolerable, so that, if man had the ability to choose, he would, like a volunteer soldier caught in a battle, renounce his original enlistment gladly. Neurotics are notoriously fond of dumb animals.

Conventional evolutionary theory does not provide for an intelligence that would direct mutations toward every-increasing self-consciousness. Isolation and inbreeding among a slightly more schizoid band would be counted upon to produce a type that would, given the chance, venture forth and shove aside less able hominids, or, later on, humans. But this cannot go on for long, unless there is a mutated element present in the germ plasm allowing ultimately the full exercise of self-awareness.

Here is an area where evolutionary thought is especially self-contradictory and, consequently, slippery and evasive. It can only get from one small change to the next but cannot get from the beginning to the end; it can explain some intra-species changes, like horse-breeding and the Beltsville turkey, but it cannot explain a major development. No known mechanism directs a long string of slight modifications in the germ plasm. Even if we were to concede that the jump from hominid to human were only apparently large but was biologically small, human genesis would admittedly be a hologenetic occurrence; when it occurred, hominid life changed drastically; it speciated.

BRAIN SPECIALIZATION

Nor can humanization wait upon a slowly evolving culture, no more than the bee was anatomically created and then evolved the basic elements of its social system over millions of years. Even though he does not draw the consequences - hologenesis - we can agree with Robin Fox when he writes: "The nature of order is part of the order of nature. It is not that man is as culture does but that culture does as man is."[88]

Recent researches into the differing behaviors prompted by the separate hemispheres of the brain can also be considered. Hominid 'X' may or may not have had a large brain before he was humanized, that is, before he became schizotypical. The fibrous conjunction (corpus callosum) bridging the left and right hemispheres of the brain may be playing an effective role in conditioning humans for schizotypical behavior, even if it is not indeed the physical location of the genetic factor that so many are searching for.

In his treatise on *The Ghost in the Machine,* [89] Arthur Koestler has placed the origins of human 'mis-behavior' in a malfunctioning relation of the limbic system to the cerebral region. The basic reptilian and mammalian control and response systems are located below and behind the cerebrum, which is grossly 'over-developed' in man. The rational and constructive inclinations of the uniquely human cerebrum, he thinks, may be frustrated all too often by the more instinctive, unconscious, and irrational animal systems. Human behavior, as a result, is prone to contradictions, rage and aggressiveness, destructiveness, and madness.

Even while admitting that a specialization is occurring here in the human central nervous system that can bring about schizoid behavior from a lack of perfect coordination, we must say that the problem is incorrectly stated and may explain why Koestler did not arrive at the focal center of human nature. The problem is not one of 'mis-behavior' but simply of behavior, both 'bad' and 'good, ' 'normal' and 'abnormal. *'Pari passu,* there is no 'malfunctioning, ' but only 'functioning. 'We do not turn off a spigot marked 'rational' and turn on the spigot labeled 'irrational.'

Once we brush aside this specious and decrepit Aristotelianism and scholasticism, Koestler's work becomes valuable. For now it becomes

---

[88] *Biosocial Anthropology.* London: Malaby, 1975, 7.

[89] New York: Macmillan, 1968.

possible to seek a mechanism of delayed instinct between the automatic and cognitive specialization of the brain, which, in conjunction with other sources of delayed, diffused, and over-loaded responses, may explain the self-awareness, existential fear, and profuse displacements of the human being.

The bilateral structure of the brain, providing two hemispheres, had been fashioned long before the advent of humans, probably one some quantavolutionary occasion between two ages. A division of functions between the hemispheres may have come only with the origination of mankind. The skullcase tends to warp to conform to the concentration of functions in the brain; and external asymmetry conforms to the internal asymmetry. Such asymmetry, implying human specialization, may characterize most or all hominids. Ornstein asserts that hemispheric specialization (asymmetry, that is) appears to be unique to humans. [90] Handedness in favor of the right hand, and language, are dominated by the left hemisphere. Asymmetry in the language region is, for instance, discoverable on the skull of "Arago XXII" coming from Tautavel, France. This specimen is classified as homo erectus and assigned an age of 450,000 years by uranium-thorium and electron-spin-resonance tests. (Source: Musée de l'Homme, Paris.)

Besides governing right hand and body movements and language, the left hemisphere is specialized in analysis and mathematical functions. It is also assertive and, in observed behavior and experiments, tends to dominate decision-making. The right hemisphere of the cerebrum initiates and supports activities of the left side of the body, and pursues non-verbal and holistic forms of thought and appraisals of experience. It is described as artistic and analogical in its ways of processing the external world for internal consumption and action. Thomas Parry has surmised that a relation exists between ancient catastrophism and a take-over of internal and external behavioral leadership by the right hemisphere of the brain on the occasion of traumatic experiences.[91]

Each hemisphere alone can convey to the whole person the possibility of physical and mental survival. Each is in constant touch with the other through the medium of the corpus callosum which carries millions of connecting links between them. The severance of this membrane has permitted direct observation of the individuality of the two hemispheres. It leaves a still "normal" person "with two separate minds, that is, with two separate spheres of consciousness."

---

[90] Robert E. Ornstein, *The Psychology of Consciousness*, San Francisco: Freeman 1972, 63.

[91] "The New Science of Immanuel Velikovsky," I *Kronos* 1, 1975, 6-7.

If the key to humanization is a general delay of instinctive response with a consequent choice-factor introduced into a wide range of behavioral decisions, then a possible source of the delay lies in the corpus callosum and/ or any drug that can inhibit the full and complete communication or near-identity of action of the two hemispheres. If, for example, fatigue and exhilaration both produce schizoid symptoms, some quantitative measure of interaction between the cerebral hemispheres may define the normal schizotypical state of the hemispheric relationship; the norm itself would be genetically and/ or socially induced on a continuing basis, providing typical human behavior. The recent association of high or uncompensated adrenalin secretion with schizophrenic symptoms suggests offering this drug as a candidate for a humanizing auxiliary.

One is inclined to distrust so simple a solution to so fundamental a problem, even after posting the usual warning signs: that the process is more complicated than it appears; that we know next to nothing about the circulation of adrenalin and other drugs with which it interacts in process; and that historical proofs of such an evolution are probably impossible.

One might as well suppose, while offering the same type of warnings, that an electrical change has brought about human behavior. If the Earth has gained charge in recent millennia, the human body may be operating in a hyper-electrical mode relative to the environment in which it evolved. This would be the case with the biosphere generally; insects, birds, and mammals are all sensitive to electromagnetic fields and changes in them. The hominid might then become the 'nervous human' who turns upon the not-quite-quantavoluted hominids and trains them to be human, meanwhile through adaptations and interbreeding creating a new race, whose, members are quantitatively distributed about the genetic norm of the 'nervous human. '

As with every significant element in the quantavolutionary theory of homo sapiens schizotypus, the hypothesis of the physiological source of humanization is put forward to orient thought and method. The theory as a whole serves to show where we can go when deprived of the assumptions of a uniformitarian external force-field of evolution and of the free, long expanses of evolutionary time.

SIGNALING HORMONES

A logical candidate for mutation and environmental transformation in the chaotic period is the endocrinal system. It is an anciently derived collection of glands, separate from but connected with the brain, the nervous system, blood pressure, metabolism, growth, sex, fear, and stress.

It discharges numerous hormones that stimulate and regulate these systems. Its main components are the pituitary gland, the pancreas, and the adrenal cortex and medulla. Lionel Tiger places "phyletically prescribed environmental boundaries" around "sociogenic processes," treating mainly of endocrinology. [92] The bio-social movement may help quantavolution much, because of the intense scrutiny it gives to the logically necessary biological and social interface where the great change of humanization had to occur.

The endocrinal system, especially the adrenal cortex, is stimulated by stress and establishes counter-stresses in the organism. For example, rats bred in the laboratory have smaller adrenal glands and less resistance to stress, fatigue, and disease than wild rats. Their thyroid glands are less active and their sex glands develop earlier and permit greater fertility. They have smaller brains, are tamer, and are more tractable.

In humans, similar differences occur between people who are stressed by the environment and those who are not. New Yorkers usually have enlarged adrenal medullas, compared with the American population at large. Paranoid and obsessive traits, involving distortions of reality, are commonly observed among persons who suffer from an excess of adrenalin either as a result of great fear and anxiety or in consequence of inadequate suppressive and discharging chemicals and mechanisms.

Schizophrenia involves at least some separation of the 'primary' self from a second self, which includes part of the self and engages in profuse identifications with the outer world. Frequently observed in mind-workers, it evidences heavy pituitary stimulation of the brain as well as insulin and adrenalin 'excesses.' The brain often becomes ungovernable owing to endocrinal disturbances. Notable, too, is the association of fear, aggressiveness, and sexuality in variations of the endocrinal system. It is then reasonable to suppose, for instance, that sexuality is determined more by the stresses of the quantavolutionary period than by the aboriginal oedipal complex or simple sexual drives.

Other modes of mutation or transformation also point to the importance of the endocrinal system in developing humanness. Solar radiation stimulates the adrenal system, both directly and indirectly. Hence, abruptly changed levels of solar and other types of extraterrestrial radiation may have prompted humanizing behavior. The types of social imprinting imposed upon the first generations of mankind and all generations since

---

[92] "Somatic Factors and Social Behavior," in R. Fox, ed., *op. cit.*, 115; E. J. W. Barrington, *et al., Hormones and Evolution*, N. Y.: Academic Press, 2 vols., 1979.

then were, so far as we can tell, the same; delusory, symbolic, obsessional, and aggressive; these are typical products of endocrinal excesses.

Finally, the obsessive will to mutate, to change one's corebeing down to the egg and sperm themselves, has been proposed by Freud as an evolutionary example of "the omnipotence of thought;" so strong a will would be more probably and capably generated in individuals who are endocrinally excited. More than by growth of the brain, therefore, the accelerated development and passover of hominid to human in a quantavolutionary period may be owed to the endocrinal system.

The hypophysis or pituitary gland excretes hormones that can arrest growth and cause dwarfism by reduced excretion, or giantism by increased excretion. An increase also probably increases the rate of insulin secretion by the pancreas. Growth hormone "directly enhances transport of at least some and perhaps all amino acids through the cell membrane to the interior of the cells."[93] It also depresses glucose utilization by the cell. The growth hormone is secreted continually from birth to death.

If the hormone reduces or perhaps delays growth, and at the same time can deprive the cell, including the brain cells, of nutrient amino-acids, and can also diminish insulin output, can it then contribute to the delay and dispersion of signals through the brain? It is conceivable; experiments can be designed to test the hypothesis.

Man is supposed to be fetalized as compared with the apes since in the adult man the size of the head and the relative proportions of its parts resemble those in juvenile apes rather than those in adult apes. Bolk speculated that fetalization may have been caused by changes in the hormonal balance in the body, especially by a decrease in the production of the anterior pituitary hormone.[94]

Dubrow has correlated growth and size of humans and many other life forms with changes in the intensity of the earth's magnetic field. We may wonder then whether an endocrinal change produced by a change in the GMF might stimulate pituitaryism and expand australopithecus to modern human proportions.

Since the left brain hemisphere is asymmetric with the right hemisphere, being larger occipitally, and this area is close to the calculating and speech centers, then a growth of the total cranium implies an important proportionate growth of this area and its special functions. That it may be more than proportionate is indicated by Dubrow's finding that the length

---

[93] A. C. Guyton, *Medical Physiology*, 3rd ed., Philadelphia: Saunders, 1966, 1040.

[94] Dobzhansky, *op. cit.*, 205; L. Bolk, *Das Problem der Menschenwerdung*, Jena: Fischer, 1926.

of the skull geographically varies inversely with the intensity of the GMF.[95] Thus humanization would accelerate. The quantavolution that split man's mind and freed it to displace copiously upon the world may thus have been influenced by a declining GMF. This 'freedom' would then take the form of the multiple selves, or poly-ego.

I have noted on occasion that drugs which are used to treat diabetes of the pituitary variety, and are intended to reduce blood glucose concentration, occasion paranoid suspiciousness as a side effect. But this and these other workings of the endocrines are puzzles within riddles: as F. Dunbar said, "There are no disorders of single endocrine glands."[96]

MUTATION

Let me consider now mutation, asking the ethologist and expert upon instinct, Tinbergen, to describe the situation:

> Present day theories of evolution consider mutations in the widest sense as the basis of all heritable change. The variability due to mutational change may show directiveness of various types, adaptive as well as non-adaptive. Adaptiveness is brought about by selection. Speciation, or the divergent evolution of populations originally belonging to one species, starts with geographical expansion of the species' range to such a degree that two or more populations of one species become reproductively isolated. The various populations thus isolated are usually slightly different in genetical make-up right from the beginning. This difference, together with the environmental differences leading to different selection pressure, account for divergent evolution of the populations which ultimately results, via the formation of geographical races, in the origin of new species, genera, and even families. Whether this 'micro-evolutionary' process is at the bottom of all evolutionary divergence, even of those often called macro-evolutionary, is a matter of disagreement. It is certain, however, that the causes of evolution can only be studied in micro-evolutionary processes.[97]

A gene is a large molecule of deoxyribonucleic acid (DNA) wound on a double helix, along which are strung in fixed order some simple chemical structures called nucleotides. Only 4 types of nucleotides are ordinarily

---

[95] A. P. Dubrow, *The Geomagnetic Field and Life,* N. Y.: Plenum, 1978. *Ibid.,* 84.

[96] *Emotions and Bodily changes,* N. Y.: Columbia U. Press, 1935, 4th ed., 1954.

[97] N. Tinbergen, *The Study of Instinct,* Oxford U. Press, 1969, 5th printing, 195.

found in one and all chromosomes but their varied arrangements establish by code the behavior to be followed by any given gene. Hence each of the supposed 50,000 genes that carry the full hereditary code of instructions has its unique code that determines its unique job.

A gene mutates, that is, changes its code and hence its 'building plan' by a disarrangement or loss or destruction of one or more nucleotides of the helix. This accident occurs when a foreign chemical or particle or charge or wave or organism enters the chromosome and its gene, with especial effects when the gene is in the process of duplicating itself. Once the gene is altered, it transmits new instructions and whatever aspect of the organism is under its command will accordingly change. Mutations may also affect the organization of genes within the chromosome, rearranging them or even rearranging chromosomes.

It can be estimated (following calculations by Wallace and Dobzhansky) that in the case of man, the number of 'spontaneous' ('natural, ' 'background') mutations that would occur for a world population of four billion people in 350 generations amounting to 10,000 years would be only around two hundred (200). Since practically all mutations are 'cosmetic, ' harmful, or lethal changes, it is embarrassing to place one's faith in mutation (at least as here construed) as the factor bringing about speciation from hominid to man. Indeed, Wallace and Dobzhansky, after presenting the negative and positive effects conclude that mutation is something to be avoided. More-over, "Lack of genetic variability for further evolution of the human species is something we need not worry about."[98] Like the last man to squeeze aboard a crowded bus, they don't want the driver to stop anymore to pick up someone else.

Here, however, we are concerned with the point of origin of the bus: presumably the change from hominid to man must be applauded. Somewhere along the way this genetic event occurred. But we can understand the plight of uniformitarian evolutionists. How many mutations are represented in the differences between hominid and homo schizo -- one, ten, fifty, one hundred, one thousand? Geneticists cannot say, because, excepting a very few cases, they do not know yet what genes control what changes to what degree. (Anthropologists, such as Washburn and Moore, in their book *From Ape to Man,* can brave the statistical jungle to extrapolate, but fail.)

If we retrocalculate the figures given above, we would have, say, a single viable mutation per ten million years. For, if the humanizing population is set at four millions instead of four billions over whatever time period is

---

[98] *Radiation, Genes, and Man,* N. Y.: Holt, Rinehart, Winston, 1959, 43.

involved, a generous estimate by conventional reckoning, then we multiply the time required for 200 mutations one thousand times, giving 10,000,000 years. If one in 200 mutations is viable, then we get a viable mutation every ten million years. But the difference between viability and the ape-to-man difference is still to be bridged. Would then 500 viable mutations be required in order to bet upon the critical change occurring? If so, this would appear to require five billion years. Fortunately, we can dispense with further arithmetic, since authorities have pronounced this to be the age of the Earth itself.

To explain the creation of man by mutation under a uniformitarian theory is thus impossible. To call in natural selection, as is usually done, does not help. For natural selection, unless it is sheerly *ad hoc* or *post hoc ergo propter hoc* reasoning, must have some genetic possibility to work with. It must depend upon mutation to begin with. One cannot assume that homo sapiens resides in 'Hominid X' like a homunculus, awaiting only isolation and inbreeding, or a shift in moisture-carrying winds, or a new supply of protein-rich alligator meat to give the creature sustenance.

If we are to use mutation theory at all, we must associate it with radionic turbulence of the most violent kind, extended over many centuries. And it is even more plausible if, to such catastrophic mutators, we add a permanent change in some atmospheric constants. And, then, too, we must continue to belabor mutation theory, for there is some deep mystery in it -- a kind of genie in the bottle, something of Lamarck's environmentalism, of Freudian psychosomatism, perhaps even of the monads with miniature universes within them, which is to say, something of a Great Intelligence. At this point in time, then, we still need mutation theory and catastrophe theory, with an open door to whatever other theory comes bearing fruit.

We receive a hint that the merger of gene theory of mutation into macro-evolution or quantavolution is possible with recent studies showing that much DNA (like much brain tissue) is surplus, seemingly unnecessary.[99] Is this material that is in readiness for recombination? Is it potentiated for organizing a general response in the event of a mutation that would otherwise be too specialized to survive in the species?

Some important areas of agreement exist concerning mutations. Genetic mutation is a change in the formation-instruction code contained in the DNA component of one or more genes of the sperm or egg. Of the estimated 50,000 genes, a mere 210 have been assigned loci in specific

---

[99] Discussed in B. Silcock, "The New Clues that Challenge Darwin," *Sunday Times of London*, Aug. 3, 1981, 13.

chromosomes.[100] The gene map is practically useless, then, in plotting the route of humanization.

New genetic instructions are carried into the fecundation of the egg, thenceforth into the embryo, the newborn, and the adult. This happens provided that survival is possible under the changed rules of growth. "Mutation" of non-genetic material whether adult or embryonic, affects only the individual and is not reproducible. Many chemicals and particles can bring mutation in this sense; but they affect individuals, not species, through cancers and abnormalities.

George Gaylord Simpson laid down a few years ago several principles that are pertinent to humanization.

> It is now known that mutations, broadly speaking, an ultimate basis of variation, are discontinuous... The somatic effects of mutations vary from great to barely perceptible, or quite likely, to unperceptible by usual methods of observation... Despite the fact that a mutation is a discrete, discontinuous event at the cellular, chromosome, or gene level, its effects are modified by interactions in the whole genetic system of an individual (oddly enough, there is no generally accepted term for that important concept). They [mutations] are also modified by varying environmental factors. The results are that for many mutations, the somatic effects in different individuals vary in an essentially continuous manner. Even an expression that is marked modification in some individuals may be only the extreme of what is a gradual sequence in the population.[101]

The whole genetic system falls into line with the mutation, so to speak. This is certainly a hologenetic effect; one wonders why "no generally accepted term for that important concept" exists. A great many features of the organism (hence species) are systematically calibrated. Still, individuals of the species, already unique, alter in unique ways as a result of the mutation. Whether the human 'big brain' evolved in one or several steps, the process was individualized so that, for instance, one person could have only half of the cranial matter of another person; further there are ethnic and sex differences in cranial size.

Simpson hesitantly comments on the likelihood of quantavolution of species:

---

[100] V. A. Mckusick and Frank A. Ruddle, "The Status of the Gene Map of the Human Chromosomes," 196 *Science* (22 April 1977), 390-405.

[101] *Op. cit,; The Major Features of Evolution*, N. Y.: Columbia U. Press, 1953.

The instantaneous origin of a new species by a single genetic event can occur but is unusual. It is practically confined to cases of increase in individual chromosome numbers happening to produce a system both viable and capable of reproduction but not capable of backbreeding into the parental population. In usual... cases distinct evolutionary change involves the increase or decrease of proportions of genetic factors in whole populations, and this is a gradual process occurring in successions of generations. The prevailing modern theories of evolution are essentially, although not dogmatically, gradualistic.[102]

No new species has been proven to form, through mutation, breeding, or otherwise, in human history. First mutations occur rarely; perhaps one in 25,000 spermatozoa or eggs possesses a gene that has been mutated. Still, with a large population over a long period of time the number of mutations will be high. Since women carry their eggs from birth, some 200 of them, an egg mutated on one occasion may be represented in a birth as much as forty years later; male sperm is wasted and renewed, millions per ejaculation, so that a mutated sperm has very little chance of being partner to a conception.

The chances for a successful mutation are so slight, and the process typically visualized by biologists for evolution of a species is so long, that many scholars have offered calculations showing the high improbability of the origin and development of species by mutation. Yet other theories have not been acceptable, except for the enlargement of the mutation-referral, or calibration process that Simpson spoke of above. Natural selection has to work only with the gene pool already available to a species and is questionable on the grounds already stated in the preceding chapters.

MAJOR FEATURES OF EVOLUTION

G. G. Simpson declares that "Mutation rate can rarely be an effectively determining factor in rate or direction of evolutionary change; this is also the conclusion of Muller..., leading student of mutation rates."[103] Mutation offers plenty of possible changes but natural selection is more important: once more we face the frustrations of evolutionary ping-pong between mutation and natural selection.

---

[102] *Ibid.*

[103] *Ibid.*

An effort was made by the geneticist Richard Goldschmidt, in 1940 to provide a new material basis for evolution.[104] He said that his lectures at Yale ought to have been called "the genetical and developmental potentialities of the organism which nature may use as materials with which to accomplish evolution." Evolution and natural selection, including the survival of the fittest, were accepted by him as facts. But, he said, selection and adaptation required "necessary hereditary variations" to work with.

So he strove to discover evidences of "macro-evolution." He showed how hereditary differences, that might have fateful consequences in appearance and behavior among species, might be attributable to certain mutations. He conceived the idea that "hopeful monsters" would be frequently generated, from among which some rare type might accomplish an evolutionary saltation. Although he could not demonstrate such directly, he conceived that novel patterning of chromosomes might instantly achieve the same effect as an accumulation of mutations, producing a new chemical system that would substantially alter an organism's appearance and behavior. So he could speak of "systemic mutation" as a complete change of the primary pattern or reaction system into a new system.

He might have added the term "hopeful scientist," to describe himself and others who were products of the "hopeful monster," homo schizo. The phrase: "To illustrate the presence and wisdom of God in the natural and moral world" meant to the naturalist, he declared, "the demonstration of law and order in his chosen field." This view is a common amnesiac sublimation of the characters of the gods Yahweh, Shiva, Zeus, and Jupiter, spreaders of chaos and lightning-like destroyers of the order of Mother Earth and Mother Nature. Perhaps if he had investigated the character of his gods, he might have truly found the means by which nature accomplished changes -- by catastrophes multiplying infinitely the mutating forces and adaptative opportunities of the world. He then would agree even more with another authority whom he supports, O. H. Schindewolf, the German paleontologist, who not only surmised macroevolution but adjudged the causes to be catastrophic and extraterrestrial, in a set of studies published between 1936 and 1963.

The earliest men were in fact "hopeful monsters" who had to believe that the gods were responsible for their sorrows, as well as their welfare, but sublimated many of the sorrows. Perhaps this is why Mircea Eliade, the hopeful scientist, must wonder why the first Greek god Ouranos was believed to have bred so many hateful monsters, his own children, whom he cast down and buried in the bowels of the Earth; Eliade may be avoiding

---

[104] *The Material Basis of Evolution*, New Haven: Yale U. Press, 1940, 3.

his own ambivalence in not answering the question that perhaps he of all scholars is best equipped to answer.

Coming closer to the key to quantavolution and macroevolution are scientists such as Dubrow, who credits sharp changes in the geomagnetic field with mass mutations leading to sudden increases in populations and systemic mutations leading to new species and genera.[105]

INTELLIGENT MUTATION AND EVOLUTIONARY SALTATIONS

That genes instruct organisms via chemo-electric code is well-known. That genes mutate occasionally has long been known. The mutation as an electro-chemical event with functional consequences is also appreciated. Puzzles remain: how, if at all, do mutating genes provide the non-random set of instructions needed to accommodate the rest of the organism to the new structure/function of the changed part? The problem is made all the more poignant by the observation that nearly all mutations are relatively "meaningless" and mostly trivial; yet a given species is integrated functionally, and differs "significantly" from another species. No gift of time, no matter how generous, nor even the bonanza of radiant catastrophes, can displace our feeling that mutations may generate "hopeful monsters," some of which survive.

A new metaphor is therefore suggested. We assume that the mutation is a changed chemical message sent by one gene to all other genes as well as to all other genes as well as to the operations which itself commands. Every gene (hence chromosome) receives, upon mutation, not only a capability to provide a new instruction but also a capability for leadership. Every gene, like Napoleon's soldiers, "carries a marshal's baton in his knapsack." When it mutates, all other genes become dedicated followers. The gene, as befits the ideal field marshal, conveys to them instructions concerning the behavior newly expected of them. They do their limited best to conform to the new order.

The gene that gives the limbs of my cat a surprising six digits orders all other genes to whom the change is relevant to provide the necessary services. Muscles, brain, blood vessels, and many other structures and functions swing into line. The cat survives and breeds its kind.

The instructions given out by the other genes that control the cat's features are contained in their programs, for apparently they have not been

---

[105] *Op. cit.*, 99.

mutated. We see at least two levels or types of changed instructions passed from a leader gene to all other genes: a) a new proportionality of structure and function which provide 'normal' individuation within limits of an ongoing species, and b) *ad hoc* accommodations in the presence of hitherto inexperienced demands. The *ad hoc* accommodations may be presumed to be quantitative or extreme deviations of the individuation code. Both types of change will persist so long as the mutated gene gives off the same signal, practically "forever." The mutation may be "deleterious," or "harmful," if the pre-existing capabilities are not flexible enough to provide holistic means of survival; on the other hand, the *ad hoc* instructions may be accommodated, and the organism survives.

Is it conceivable that the genes carry design accommodations for every successful macromutation? If so, where do they originate? Suppose that, in the beginning of life forms, each gene is possessed of designs that can cope with every form from an amoebae to a whale (this is, of course, not a new idea). Given a certain chemical stimulus, it will produce its part of the structure and function of any species known up to the present and many more. There is no logical reason why an individual gene capability of a bacterium of 2^{2000} combinations cannot foreshadow all life forms that have developed. The gene's speciated repertoire of designs presumably has limits. Indeed, such limits are commonly defined in the course of reciting the similarities among all living forms. They are further defined in the course of classifying phyla, orders, and families. Then, should a changing gene stretch the führerprinzip too far, asking, for example, that feathers be provided for a whale, the followers, the other genes, cannot find the requisite function among their repertoire of attainable specifications, and the animal will usually die.

But suppose that my cat bears kittens with flipper-like limbs. The mutated gene passes its changed chemical messages to its cohorts and the necessary changes are made, well within limits, except that the little beasts cannot walk very well. They might swim and, if introduced to a body of water, do so. The cohorts work hard now to accommodate: eye muscles tighten; muscles bulge at the flipper joints; oil gathers heavily at the skin pores; the body becomes rotund for insulation and floating; the taste buds are alerted to watery savors; the lungs expand for searching and diving underwater; and so on.

The young cats are not equally flexible and they lack parental instruction, but perhaps one or two survive and go ashore, mainly to procreate and give birth, cautious and suspicious of land forms, abandoning their un-mutated kittens and carrying their mutants back into the water. Since my cat is a mixture of Siamese, Persian and Mediterranean alley-cat, its kittens and their kittens will afford numerous possibilities for immediate "natural selection." They will compete adequately with beaver, muskrats, otters, duck-billed platypus, seals, and sea-lions, and will supply prey for the few large carnivores of the sea and food for smaller marine animals with their carcasses. They may live and hunt in gangs. If my cat had given birth to all of this in secret, I would be surprised by a new order of beasts when, a few years later, I would be swimming.

Should, in the course of events, a member of the new species be mutated, a new gene would probably become the leader. Now gigantism is the order of the day, from among the dead-born emerge two double-sized kittens, which grow to quadruple-sized adulthood. A new instruction would have been dispatched to all its genes, which would have been received and interpreted on the basis of previously existing instructions, not for "Cat" but "Cat I."

The process of fixing the next species, "Cat II," would be analogous to the earlier one. Cat II genes would be centered around the Cat plus Cat I norm. Their limits of deviation presumably would remain those of Cat plus Cat I. That is, they had inherited Cat I's new instruction. The new chemical instruction would build upon it; it would only order research of the repertoire to its limits to abide the new order. So, in swimming around a decade later, I might receive an even greater surprise.

Examining the gene structure of Cat II progeny, one would find all of the instructions implanted in the primordial form of life, Amoebae I (or even, in fact, its predecessors, that were locked into it), together with every mutation (or new command) ever imparted to Cat II ancestors. Missing would be only the changed genetic capabilities afforded species that have branched off of its line since the beginning of life. The whale would be denied feathers. In this sense, "ontogeny recapitulates phylogeny." Under such conditions, the number of successful mutations from the primordial form might have been far fewer than is generally believed, perhaps less than a hundred for the generally of species, and under a dozen for the particular species.

Consequently, mutations can be conceived to cause very little or very great changes in the structure and functions of a species. Further, mutations are considered statistical, that is, indeterminate increments contained in a limited number of commands. They are, of course, not models of the form-to-be. Their "intelligence" consists in their primordial ability to induce coordinated shifts of behavior in non-mutated genes.

By implication, "important" changes occur by saltations, as quantavolutions. In environments that provide mutational possibilities, radically different forms can emerge quickly, propagate abundantly, and branch quickly again. Long-time durations are of little importance; whether they occur or not is immaterial.

The theory here is so simple that it may be merely a metaphor. It need not be justified by elaborate mathematical calculations. It preserves most of the general observations of Lamarck, C. Darwin, St. Hilaire, Mendel, Dobzhansky, Watson and many another geneticist, it can cope with paleontology and genetic engineering without strain. It suggests, among other things, that, in principle but against great odds, preexisting ancestral species can be recreated, and that the creation of future "major" life forms is within sight, perhaps at the level of probability of controlled nuclear fusion.

EXTERNAL PRODUCERS OF MUTATION

The prevailing evolutionary theory, "The Modern Synthesis" has looked to point mutation within structural genes as causing individual variability, which is ultimately carried into a population where it comes to be a dominant trait. A species change is thought to occur by gradual accumulation of small differences. Isolation and small numbers promote the change. Subsequently, the new species diffuses. Long time intervals are admittedly required. Transitional forms, which should be abundant among fossils, are rarely discoverable, and never incontrovertibly accepted as such. The fossil record appears to be a representation of quantavolutions, not incrementalism. It is suppressed, however, by an ideology of uniformitarian evolution.

Even when "the Modern Synthesis" is attacked, as it was recently in a conference of geneticists and paleontologists, the challengers, advocates of 'macroevolution' or 'punctuated equilibrium,' (our 'quantavolution') appear to stay within the boxing ring outlined by an assumed speciation: 'what happens in speciation? ', not 'what causes speciation? ' A rapid speciation, even to the challengers, is one "taking place over, say, 50,000 years, but that is an instant compared with the 5 or 10 million years that most species exist."[106] Even so, it would be far longer than necessary to change Hominid 'X' into homo sapiens schizotypus, if the modifications which I suggested above were sufficient to make the main differences between the two species. Once the viable combination is struck, the speciation occurs instantly.

Furthermore, with normally prevailing rates of mutation, speciation is unlikely under either the Modern Synthesis or the 'punctuated equilibrium' theory. It is striking that the aforesaid conference did not take up the

---

[106] Lewin, *op. cit.*, 883.

question of the possible role of cosmic or space environmental change. Writing in 1980, a group of scientists claimed that a major extraterrestrial impact on Earth ended the Cretaceous 'reptilian' period and inaugurated the Tertiary mammalian period at which time, quoting D. A. Russell, "no terrestrial vertebrate heavier than about 25 kg is known to have survived," and the food chain was completely disrupted for many years by other biosphere extinctions and reductions. Further, "there have been five such extinctions since the end of the Precambrian," bringing us back to the beginning of life.[107]

Schindewolf, Salop, and a number of other scholars, whether in the close fields of genetics, geology and paleontology, or in the general field of catastrophism, have brought forward volumes of material to support the likelihood of mutation-causing disasters. Probably the 'earth-bound' specialists are waiting for a green light from the astronomical establishment. Meanwhile pressure mounts from the earthlings and the general catastrophists. *Nature* magazine, for example, carried in one issue (May 22, 1980, Vol. 285) three articles on catastrophes at the Cretaceous-Tertiary boundary.

Not only mutations, but all other factors in speed-up of genetic change are provided by natural catastrophes -- isolation, adaptation, and extinction of competing species. Thus we hear Simpson say that "The chance of fixation of a favorable mutation may be considerably larger by accident of sampling in a small population than by selection in a large population..."[108] Catastrophes therefore simulate in quick time the supposed effects of natural selection. If man has been humanized within the past 100,000 years, or even within the past million years, actually at any age boundary, even granting the dubious long-time reckoning, he probably was humanized by catastrophe.

Here the quantavolutionary model diverges from the evolutionary model most emphatically. In order to enhance the chances of a viable speciation by mutation, a heavy bombardment of particles is required such as has not been experienced in history; a radiation storm is called for. Such storms must have existed on numerous occasions in recent prehistory, if the evidence assembled in my *Chaos and Creation* is accepted. They ionized thoroughly the environment by interrupting, deflecting, and reversing the electromagnetic field of the Earth, by mega-lightning electrical discharges

[107] Luis W. Alvarez, W. Alvarez, Frank Asaro, Helen V. Michel, "Extraterrestrial Cause for the Cretaceous-Tertiary Extinction," 208 *Science* 4448 (6 June 1980), 1095-1108. 1107; Russell, *Episodes* 1979 No 4, 1979, 21. *Cf.* Otto H. Schindewolf, "Neocatastrophism?" (trans. V. Axel Firsoff), 2 *Catastrophist Geology* 2 (Dec. 1977), 9-21.

[108] *Op. cit.*

upon the near encounter of bodies in space, upon the occurrence of great potential differences between space and Earth, and by removing cloud canopies and transforming the gaseous composition of the atmosphere. Meteoric pass-throughs collisions would have occurred. The Sun would be stimulated to hyper-activity. The electrical and atomic state of every organism and rock would be altered.

The radiating effect of one meteor or comet of small size gliding through the atmosphere is heavier than that of a large cluster of hydrogen bombs because of its great heat, well over 2000° C, over a long trajectory, the wide distribution of fall-out, and its possible final explosion at a great speed of many kilometers per second. A single such passage, of which there would have been many, should produce millions of mutations in the biosphere generally. A large explosion creates a catastrophic tube from the upper mantle into outer space, in and around which many millions of combinations of electrical, chemical, physical, material, thermal, and pressure events take place. Paleontologist D. J. McLaren had events such as these in mind when, in a presidential address to his colleagues, he reviewed the evidence of the wholesale extinction of species. After describing the effects of a large-body collision, he remarked: "This will do."[109] Yet, it is not only extinction that occurs, but also speciation.

As soon as they will grant the occurrence of extraterrestrially caused disaster, paleontologists will arrive at a public agreement in favor of quantavolution. Essentially this would include first that the species have been created and exterminated in waves. The waves will probably be fixed chronologically at the passages between the conventionally named periods -- such as between the Pliocene and Pleistocene. Thirdly, they will probably settle upon radiation storms (or, for stretched-out changing, new atmospheric constants) as the principal force bringing in the great changes. They may well decide that these radiation storms are connected with cosmic explosions and encounters. Finally, they may, with greatest reluctance, turn to a shorter time-scale for measuring the succession of events in natural history. For radiation storms and geological disasters not only mutate and exterminate species; they also invalidate methods of dating that assume a constant chemical and geophysical environment.

The time of man and protoman now includes a Holocene that impinges upon the Pleistocene that is moving back in turn into the old Pliocene. I have already noted that anthropologists believe that they are finding modern types of *homo* in early Pleistocene (once Pliocene) times. Whether the advent of homo sapiens should be set in these times or in the early Holocene

---

[109] Quoted in Robert Bass, "Did Worlds Collide?" 4 *Pensée* (Fall, 1974) 8.

depends largely upon whether one adopts a long-time or short-time chronology. The change from hominid to *homo* was not anatomically or physiologically spectacular. Australopithecine and sinanthropus, if they lived alongside each other, probably lived in the time of proto-modern man as well.

Ericson declares our thesis in the title to his study, "Extinctions and Evolutionary Changes in Microfossils Clearly Define the Abrupt Onset of the Pleistocene."[110] Now I report what Salop writes:

> At the end of the Pliocene, some 3 M Y ago, is the last great revolutionary limit-line in the history of life. The first ancestors of Man appeared and an essentially new epoch started, the Anthropogene or Psychozoic. All other changes in the organic world, however important, seem of minor significance in comparison with that event. The animal kingdom of today originated, broadly speaking, also at that time, not counting the extinction of large animals in the second half of the Pleistocene thought to be largely caused by the activity of Man.[111]

This last explanation, involving man, is not acceptable; all species were under extinction stress in both Pleistocene and Holocene, including man. Moreover Salop limits the causes unduly. He comments, "Growth of the solar constant by one percent results in an insignificant rise of temperature of the troposphere, but the UV radiation multiplies 100,000 times." The ozone shield would capture most of this if it were as strong as today. Recent planetary, cometary, and meteoroid catastrophes, which are more probable but are not discussed here by Salop, would engender infinitely greater radiation storms.[112]

Most large mammal species were wiped out in the late Pleistocene, 70% by one estimate, in ways that would imply worldwide atmospheric revolution, as with the mammoth. The quantavolution of hominid into homo sapiens could have occurred on one of numerous occasions. Given the lesser resistance of the mammals and man to radiation effects, and granted findings such as Ericson's and Salop's, there is further reason to hypothesize the mutation and drastic adaptation of humans.

If the proto-men (the Hominid 'X') of this era were spread over at least the Afro-Asian world, some estimates, no matter that they must be highly

[110] David B. Ericson, 139 *Science* 3356, Feb. 22, 1963.

[111] L. G. Salop, "Glaciations, Biologic Crisis and Supernovae," 2 *Catastrophist Geology* 2 (December 1977) 22- 41; *cf,* Martin, P. S. and H. E. Wright eds., *Pleistocene Extinctions,* 1968.

[112] See A. De Grazia and E. R. Milton, *Solaria Binaria,* Princeton, N. J.: Metron Publ., 1983, for discussion.

speculative, are in order. The creatures must have been numerous. In a world of ten million hominids (30 per 100 square miles) and during a thousand years of one or more ionizing forces, whether continuous or intermittent, five million females would be subjected to radiation. Their eggs would be present and available for mutation for a life-span, say, of forty years. Assuming that females averaged a pregnancy once every two years, that their life spans averaged twenty years of child-bearing, and that a radiation storm environment persisted in which one of twelve fertilized eggs had been mutated, then some 1.85 billion mutated births would occur in the one millennium. Mutated sperms might raise the number to three billions.

Of these three billions, from 300 to 3 millions might be beneficial or inconsequential, guaranteeing at the least an average chance of physiological survival beyond infancy. One must not neglect the chance that two mutants would interbreed, making possible combinations of genes, or a new total configuration. If systemic mutation were admitted to be possible, then too the chances of an emergent human would be increased.[113]

The numerous high energy forces would have had enormous effect upon the ecology and mankind. Not only would they cause destruction on a grand scale; they would affect the mind of future generations in many ways -- genetically, by imprinting, by social indoctrination through story, a custom and institutions. The beginnings of mankind had to be associated with fearful happenings, as Nietzsche, Freud, T. Reik, I. Velikovsky, and of course all sacred historians have declared. Much was forgotten and distorted.

No one has detailed particular disasters and their human effects as well as Velikovsky. Still, it is not alone, as Velikovsky puts it, that mankind has never recovered from the terrors of catastrophe: homo sapiens schizotypus did not in fact exist before the terrible times. Mankind was born out of catastrophe and achieved his delusionary schizoid human nature out of catastrophes; and he can never be anything but the kind of creature that went through those special overwhelming experiences. Humanity was created during a natural reign of terror.

---

[113] I am using the kind of reasoning about genetic change over time employed by Simpson (1953), 109-10, on the horse. He estimates 300 effective new steps were needed over 15m/ y with a mutation rate of .000 001 and no systemic mutations, or macromutations, which, he says, are unknown. See also J. B. S. Haldane's approach, "Natural Selection," 101-49, in P. R. Bell, ed., *Darwin's Biological Work: Some Aspects Reconsidered,* Cambridge (Eng.) U. Press, 1959. See also Wallace and Dobzhansky, *op. cit.*

VIRAL MUTATION

Quite recently, the role of viruses in genetic change has come to be recognized. Viral storms might accompany the large-scale penetration of the atmosphere by exploded material from extraterrestrial events. Various ancient myths report such occurrences. Apollo was the Greek god of plagues and arrows; he was a sky god and not the sun, as later writers supposed. Recently, Hoyle and Wickramasinghe have, in their book *Life Cloud,* proposed that early life forms were deposited on Earth by cometary fall-out. In *Disease From Space* (1979) they also claim space dust as the carrier of plagues to Earth. Their proposed investigations went unsupported by the grants authorities.[114] The search for viruses in meteorites and Martian soil samples in proceeding.

Deeply buried viruses might also be exposed and some of them mutated by large-scale earth upheavals. Large explosions can create drifting material that will disseminate both crystallized and already activated viruses in similar fall-outs, and with similar genetic results. Hope-Simpson in 1978 reported that the last six peaks of sunspots coincided with pandemic influenza, possibly from increased cosmic radiation which mutated existing viruses, enabling them to evade human immunities.[115] Not to be ignored, therefore, is the chain reaction of a virus, a viral mutation, and a human mutated by a virus. Again, the likelihood of successful mutation is small but the participating organisms are exceedingly numerous.

We bear in mind the theory, advanced elsewhere in my studies, that several solar system bodies disintegrated during the past 14,000 years. One or more were probably carrying life forms. Viruses might persist for some years, possibly thousands, prior to their extinction, in a permanently hostile environment, and hence, while scarcely detectable today in meteorites or direct planetary sampling, would have been aboard their exploded vehicles in ancient strikes against the Earth.

PSYCHOSOMATIC GENETICS

Still another means for achieving humanization, and also mutational, would be the psychosomatic conversion of genes. For a time, the idea fascinated Freud and Ferenczi. They were influenced by Lamarck's theory

---

[114] *London Times,* Lit. Supp. April 14, 1978. *Life Cloud* (N. Y.: Harper and Row, 1978).

[115] R. E. Hope-Simpson, "Sunspots and Flu; A Correlation," 275 *Nature* (1978), 86. H. Hoaglund discusses "Some Biochemical Considerations of Time," in J. T. Fraser, ed., *The Voices of Time,* (N. Y.: Braziller, 1966), including oxygen consumption and slowing of time, and deep freezing and time slowdown of virus (325-9).

of the inheritance of acquired characteristics. On October 5, 1917 Freud wrote to Karl Abraham to this effect, saying, "Its essential content is that the omnipotence of thoughts was once a reality." When Abraham responded that he had not heard of the idea, Freud wrote that it would complete the theory of psychoanalysis by providing a theory of change through an "endoplastic" adaptation of one's own body.

> Our intention is to place Lamarck entirely on our basis and to show that this 'need' which creates and transforms organs is nothing other than the power of unconscious ideas over the body, of which we see relics in Hysteria: in short, the 'omnipotence of thoughts.' Purpose and usefulness would then be explained psychoanalytically; it would be the completion of pychoanalysis. Two great principles of change or progress would emerge: one through one's own body, and a later (heteroplastic) one through transmuting the outer world.[116]

Even before it was realized how minute was the probability of successful genetic mutation, Freud, like many another thoughtful person, like the theologians, like even Alfred Wallace and Lyell (until his old age), could not accept the piecemeal elaboration of homo sapiens according to the uniformitarian Darwinian model. With scientific catastrophism in disrepute and obloquy, they could not imagine an appropriate environmental stimulus to change.

The theory is not beyond discussion. Presumably the hominid bearer of sperms or eggs would be so drastically affected by environmental turbulence that it would will a chemical mutation upon them. Practically every tissue and organ of the body has been shown to be capable of physical change, usually deleterious, when an obsessed person focuses intense and prolonged attention upon the soma. The genetic material cannot logically be exempted from the obsessive influence; both point and systemic mutation could then occur.

The ability, conscious and/or unconscious, to engender fully intense and prolonged neuro-chemical and/or electrical energy, and to focus it upon a given tissue or organ, is given to few persons in these times. It might more frequently emerge when the environment is heavily agitated and the collectivity reflects this agitation and inspires a response among its members. That is, the alteration of the race by willing a genetic change might have occurred in the creative years of mankind. This would be a true

---

[116] Ernest Jones, *The Life and Works of Sigmund Freud*, N.Y.: Basic Books, III, 312, 341.

mutation, inasmuch as a chemical intervention or electrical impulse affecting the genes is postulated.

The psychosomatic model has a low probability. Although the terrorized hominid woman may have had the most intense desire, conscious and unconscious, to change her offsprings, how could she have known how to target the eggs in her womb? Can terror act as a chemical bullet directed at the eggs? We have noted that Teilhard de Chardin and the 'school' of directed evolution also have found it necessary to premise an inherent motivation towards progressive biological change, to go along with transmutation. Psychosomatism unconsciously targets an organ. Physical stress and psychic stress both can affect the heart, for instance. And our culture tells us: 'Don't give your dear father a heart attack by your evil conduct.' Further, there is a lore of affecting the unborn child. And witch-doctors may sometimes pretend to know how to affect one's enemies with psychic heart attacks and psychic damage to unborn children.

Psychosomatism, we can affirm, performs the seeming miraculous. But we prefer to believe here that psychosomatism is the cultural product of the already humanized homo schizo. It is an irresistible path that the fear arising out of the split-ego and instinct-delay points out to the human being. In one report, which unfortunately I have lost, the women of a tribal group are apparently capable of controlling their own fertilization by 'willpower; ' this is, if true, a possible effect on the germ plasm or on the fluids or musculature of the reproductive organs.

Freud, and Jung, also believed in "phylogenetically inherited material" but could never describe precisely its brainwork. The evidence is that certain common symbols are not learned, nor 'classical' phobias, nor the oedipal complex, nor some other symbols and practices. The human inherits not only predispositions, but even subject-matter and memory traces. Homo schizo has a natural cultural output: so goes the contention. But we can postpone this matter until a later chapter.

Freud and his associates could not come to close grips with psychosomatic humanization; the chemistry, biology and neurology were not available, then. They may not be now. Freud's reconstruction of the origin of conscience suffers from such basic flaws that one marvels at even the limited acceptance granted it. He should have worked instead upon his psychosomatic theory of mutation. He declares that, in an early family of homo sapiens, the sons, sexually covetous of their mother and other females, killed their father and ate him; ever since this significant incident occurred, a sense of guilt for the action has been transmitted through the mnemonic generations.

Inasmuch as ordinary observations of primates and other mammals reveal the dispossession of the aging and weakening "bull males" in families and hordes, with regard to a full range of values, including the sexual, it is presumptuous to build a specifically human trait upon the assumed killing and deduce therefrom some of the most important qualities of human behavior such as guilt, conscience, totem and taboo, religion, and civilization. Unless, of course, we are dealing with an animal already so advanced in the preparation of conscience, that the concoction of new provocation would hardly be necessary. It is much more likely that the ascription of morality to events such as the reformation of sexual power in a group is attributable to a "higher morality" - the instinct-delay fear - that gives in the process of its sublimation and rationalization direction to all aspects of life.

AN ATMOSPHERIC TRANSFORMATION

Geoffroy Saint-Hilaire (1772-1844) advanced two important ideas. One was that of saltation, the leap from one species to another: "the first bird hatched from the egg of a reptile;" the second was that atmospheric changes and other environmental changes bring about speciation, particularly those "respiratory fluids," which "sharply and strongly modify" animal forms.[117] His treatment was cursory and unconvincing.

But today it is more apparent that atmospheric reactions are an important factor in behavior. They might be an alternative or a supplement to genetic mutation in transforming mankind. In this case, Hominids 'X' are presumed to have an already existing genetic capability of becoming human. They are genetically preadapted to quantavolution. This genetic capability is not exercised in the hominid condition because the atmosphere contains a 'hominid mixture,' not a 'human' standard. The oxygen may have been more or less ionized than it is today, for example. The atmosphere may now be heavier (or lighter) in solar or cosmic rays, certain gases, and other chemical elements affecting biological behavior. Might some of these conditions alter human conduct?

The evidence is strong that some or all humans would be affected. Prevalence of unusual gases and metals in the workplace affect workers with psychiatric symptoms, even though they spend only a few hours their daily.

---

[117] "Influence du monde ambiant pour modifier les formes animales," *Mem. de l' Acad. des Sciences,* XII 91833) 63, quoted in H. F. Osborn, *From the Greeks to Darwin,* N. Y.: Scribner's, 1894, 199.

One can surmise from this fact that an enduring day-around condition would bring about shortly a different norm of human mentation and behavior.

In such cases, the changed constant would affect proto-humans in a number of places around the world and humanization would be a worldwide phenomenon of the age. Although I feel that such changed constants have affected human history, I doubt that they alone could have accounted for the emergence of homo schizo. Therefore, I follow generally the model of a single-shot mutation in humanization. Some cultural science support for this position will be cited in the chapters to come, diffusion of basic culture from a single point of origin, for instance.

D. W. Patten has offered, as a geologist and creationist, several hypotheses on atmospheric acquisitions from outer planets, especially affecting the ozone and the nitrogen content of the air, which would then alter the chemistry of growth and longevity. He halts at this point. [118] Temporary or permanent alterations in the gaseous and ion composition of the air could potentiate an already existing physiology, especially via the endocrinal glands and hormonal system. Both the solar and cosmic 'constants' were inconstant during much of the primeval period of humankind; even lately, though respecting smaller deviations, the inconstancy of the solar and galactic winds has come under study.

External events can introduce continuous and to some extent permanent changes (operating as a new constant), if the events and the conditions they bring about persist. So long as heavy noise, air pollution, rapid movement, and other high-stress life conditions of New Yorkers are constant, New Yorkers will tend to have swollen adrenals. Or, so long as the proportion of oxygen in the air of the High Andes is relatively low, the people there will have unusually developed lungs. A connection of the endocrinal system with megavitamin therapy has registered effects upon schizophrenia through facilitating the physiological discharge of adrenalin.

A diminished oxygen supply or incompatibility of oxygen type in the atmosphere may introduce schizoid symptoms to some part of the population. The brain needs oxygen not only to survive but to energize neuro-transmissions throughout its domain. In schizophrenics the oxygen level in the brain is sharply lower than normal. Further, frontal lobe brain activity is low. Thought dissociation may be produced by oxygen deficiency in the frontal lobe.

A radiation storm; a material fall-out; a sweep-out or in-take of atmosphere in transactions with extraterrestrial bodies; intense electrical

---

[118] *The Biblical Flood and the Ice Epoch,* Seattle: Pacific Meridian, 1966.

storms; and the dropping of canopies (opening of skies) can drastically reform the atmosphere. They might change atmospheric constants abruptly or over a period of time. The new atmosphere forces upon the hominids a new 'norm' of response. The new norm is, at least among some individuals, within the range of genetic capability. The adaptible survivors behave according to the new norm, which is to say that they now behave as "humans."

The reconstructed atmospheric constant may affect most importantly the fetal environment of the humans-to-be. This happens when the new chemicals in the air find their way into the hormonal food supply of the fetus. And/ or the new constant presents its demands for changed physiology and behavior upon the infant after birth. Man, and all life, lives off a radiation diet that is generally unperceived. Even today, delicate scientific instruments are required to detect radiation, and the symptomology of radiation poisoning is not very clear, or where clear does not readily name its precise cause.

The atmosphere of chaos was a mutator. The sun of the later Solarian Age may not have been. Nevertheless, the finally settled atmosphere has played a role in humanization. Legends around the world speak of a primordial cloudy sky. The opening of the skies would increase radioactive influences from perhaps still nearby and hot planetary bodies, and also and especially from the sun.

Exposure to helio-radiation (including ultra-violet rays) generally increases physical resistance, relieves arthritic and muscular pain, lends a feeling of well-being, stimulates ergosterol and hence Vitamin D production, counteracts rickets and respiratory disease, and kills bacteria and fungi of the skin. It promotes the healing of wounds and athletic performance; it increases the rate of basal metabolism. All of these occur at the price of occasional skin cancers, and possibly of still unknown deep changes.[119] Although they would contribute to a higher general level of health and activity, they would not create the human. Larger events are required.

The Earth's geomagnetic field has come under intense study in the past few years, because evidence now available points to reversals in the past. Whether the field has reversed quickly and often, as quantavolutionists believe, or gradually and rarely, as evolutionists think, a reversal of the North Magnetic Pole introduces as interval during which cosmic rays can descend upon the Earth unhindered and bring about mutations in great numbers.

---

[119] S. H. Licht, ed., *Therapeutic Electricity and Ultraviolet Radiation*, E. Licht, New Haven, 1967.

Some studies have indicated a coincidence of reversals with waves of biosphere extinction. B. Heezen, pioneer oceanographer (for it is on the rocks of the ocean bottom that magnetism can be most readily traced), has speculated that the last reversals was before the time of man. However, the time of man has been pushed back well beyond this period in conventional theory, and in quantavolutionary theory the times of the last several reversals are well within the human span, one having occurred in the eight century B. C. according to an examination of the orientation of iron particles in pottery of that age.[120] Yet another reversal is said to have occurred around four to six thousand years ago in connection with large biosphere and natural destruction.[121]

Furthermore, the geomagnetic field (GMF) is declining slowly. I have already introduced the work of Dubrow on the subject. If the decline has been exponential from some past peak, as I believe and will be discussing with Earl R. Milton in a forthcoming book, then the hominid was subjected to a sharply different paleomagnetic field. So we must ask ourselves whether the relaxed grip of the electromagnetic field disorganized the hominid brain and in effect created homo schizo. For he would be presented with an intellectual freedom in the form of a bewildering number of options for action instead of the more closed system of stimulus-response accorded Hominid 'X. ' The 'constant' is still changing, but slowly, today. Still the frequency of heart attacks has been convincingly associated with internationally collected measurements of geomagnetic activity as registered by magnetometers.

It may be possible, too, that many animals, including especially the primates, acquired a loosened behavioral potential at the same time, in the same way. Relieved of the heavier GMF, the minute electrical charges that operate the central nervous system may have stepped up their activity, relatively speaking, and, crowding the access points, delayed instinctive reactions and promoted displacements. An electric shock, administered experimentally or therapeutically (at this supposed new level of the human mind), provokes mental activity (mania), hallucinations, and amnesia, while reducing depression and anxiety. ECS [Electroconvulsive shock] leaves a permanent change in brain excitability.

That a marked change in the Earth's electrical field would have affected the human brain is not difficult to accept. We have mentioned that much testimony on a primordial canopy of clouds exists, at the time of the first

---

[120] Velikovsky describes this work of Mercanton and Folgheraiter in *Earth in Upheaval*, N. Y.: Doubleday, 1955, 146-7.

[121] Dubrow, *op. cit.* 84.

god Uranus (known by many names.) [122] The sky cover was probably removed in the time of human creation. The results would include a new and constant heavy bombardment of the biosphere with cosmic and solar particles. What legends frequently describe as the primordial chaos could have been a combination of actual celestial turbulence, ground bombardment, and mass biosphere mutations and extinctions, associated with the shock of being transmuted from hominid to homo. The Hebrew *Genesis* is by no means unique in referring to this concatenation of events.

Nor does this conclude speculation about the possibilities of the ancient skies. If large bodies transacted in close encounter or collision with Earth, as is argued elsewhere in the *Quantavolution Series*, large electrical charges would be exchanged between the bodies. The Earth could either lose or gain immense charges, sufficient to affect deeply the human nervous system. Then the proto-human must cope either with an enhanced or lesser charge on the Earth's surface or in the atmosphere, either as a sudden terminator event or as a new constant or both.

At this point of the discussion, the multiplicity of possibilities begins to bewilder and I would, if I could, sing the praises of "Occam's razor." Would the hominid mind split and develop instinct-delay and the poly-ego from any one or all of these possibilities? Or would man becomes stupefied, more hominidal, instead of electrified, confused, and energized? Reasoning *ex post facto*, which is to say, begging the question, I shall have to say that since he became the latter, whatever happened, even combinations of opposites, worked to the same end of instinct-delay and poly-ego problems.

SOCIAL IMPRINTING

In Seneca's ancient tragic drama, *Thyestes*, the chorus chants of the shocking fiery passage of Phaeton in his solar chariot, when each and every constellation deviated:

> This is the fear, the fear that knocks at the heart That the whole world is now to fall in the ruin Which Fate foretells; that Chaos will come again To bury the world of gods and men; that Nature A second time will wipe out all the lands That cover the earth and the seas that lie around them And all the stars that scatter their bright lights Across the universe.[123]

---

[122] Isaac Vail, *Selected Works*, Annular Publications, Santa Barbara, Calif., reprinted 1972.

[123] In *Four Tragedies and Octavia*, E. F. Wartling, trans., Baltimore: Penguin, 1966, 81.

A fifth means of transforming hominid into human nature might be by the social imprinting of shock upon the individual. The hominids again afford the basic genetic capability and a pre-adapted habitat. In this case, however, natural disasters inflict shocks upon the hominid beyond its 'normal' tolerances of stimulation. The shocks in themselves are the grossly exaggerated homologues of the shocks of 'normal' existence.

They take the form of a celestial scene inhabited by new symbolic references and other mind-openers; of terrorizing high-energy expressions including spectres and pandemonium; of crushing and effacing effects that are prolonged and of high intensity; of the ranging of the natural elements.

The shocks are so traumatic that the victims adopt response behaviors that become patterned as the essence of human nature. The traumatized catastrophical survivors retain the memories, but distort and use them in ways that are typically human. Most importantly, they devise in the very process of their own creation the social means of perpetuating their own changed mentalities and behavior. Human nature is then and thereby guaranteed by a collectivity of humans formed into a group or society. The memorial generations transmit and adapt new traumatic and 'normal' tribulations to the fixated human nature.

In explaining the development of the human mind in relation to the catastrophes of Venus and Mars in the period 1453 to 687 B. C. Velikovsky pushes beyond Nietzsche, Freud, Jung, and Eliade with the concepts of collective amnesia and aggression.[124] Mankind is destructively aggressive as a result of suppressing its memories of natural disasters. "The inability to accept the catastrophic past is the source of man's aggression... Freud did not come to understand the true nature of the Great Trauma - born in the Theogony or battle of the planetary gods with our Earth, brought more than once to the brink of destruction - which was the fate of Mercury, Mars, and Moon. Freud died in exile from his home, when a crazed worshiper of Wotan was preparing another Götterdämmerung."[125]

The view which I am setting forth embraces this criticism of Freud and the concepts of collective amnesia or repression concerning catastrophes. Also, aggression is to be correlated with this suppression, and the techniques of aggression are in a direct sense analogized unconsciously and consciously to events witnessed in the sky. Nevertheless I perceive social imprinting as

[124]"Cultural Amnesia" in Earl R. Milton, ed. *Recollections of a Fallen Sky,* Princeton: Metron, 1978, 21-30, 26-7; *Mankind in Amnesia,* New York: Doubleday, 1982.

[125] William Mullen, "Schizophrenia and the Fear of World Destruction," I *Kronos* (Spring, 1975), 70.

at best an auxiliary source of human nature, an intensifier, which itself needs to be intensified from time to time by fresh natural (or man-made) catastrophe.

The Middle Bronze age civilizations, 3500 years ago, whose trials Velikovsky describes so vividly, were pre-adapted to catastrophes; their societies behaved in ways already learned, and with institutions inherited from prior disasters. Ultimately, though, with the earliest disasters, a physiological change had to take precedence. Even in the genetic humanization of man, catastrophe was an on-looker, carrier, and psychological and cultural reinforcer of gene-fracturing elements.

John V. Myers and Warner B. Sizemore declare "that the disintegration of *objective* reality during cosmic catastrophe could produce *subjective* states similar to those of schizophrenia, and that the disintegration of subjective reality in the schizophrenic is accompanied by visions of cosmic catastrophe."[126] I argue that the reality recognized by the first human was catastrophic and his mind was as well. There was never -- and here I think we diverge from a common view of Velikovsky and a great many others, including conventional long-term evolutionists - a clean minded, rational evolved human whose mind was 'blown' by catastrophic experiences: the recurrent disasters *proved* to homo schizo that his vision of the world was correct!

THE SUMMARY MECHANICS

It is perhaps apparent to the reader by now that I prefer, as a 'holding position, ' a complicated mix of several means of humanization, altogether happening within a very short period of time. The mutation of an individual hominid is given prominence generally in the scenarios to come. But it is not difficult to switch from the one to the other, or to stress a combination. The changed atmospheric constant as the mode of humanization has the value of inherent continuity, and is as efficient as genetic mutation in explaining generational inheritance; also it permits humanization to occur simultaneously among many hominids at the same time, in the same month or a few years. We might begin a search for humanizing mechanisms that are present in the modern atmosphere but would not have been present in an atmosphere in which hominids could thrive.

The branches of the human race have changed in some respects, mostly cosmetically, since the cosmic beginnings of homo schizo. But the basic

---

[126] I *Kronos* (1975) 70.

ways of behaving as human were determined in the midst of great crises: the interruption of the Earth's motions, the loss of electrical charge, the dropping of the immense cloud canopies in deluges, and the first openings of the sky. An allotment of a thousand years would have been sufficient for these tremendous experiences to bring about humanization.

Even while mutations were abundantly occurring among all species, a single group of hominids, largely potentiated as humans beforehand, in distress and in terror, would find amongst themselves individuals of flexible, if erratic, genetic constitution, who were capable of expanded symbolic behavior and signaling various interpretations of the new giant forces of the environment.

The same group would become capable of managing its newly installed communication system, and then lend its cooperative forces to the evolving interpretation of the universe, the aboriginal cosmology. The group would be driven to adopt the new system even before all of its members shared the mutant genes. In the endeavor to ease their pains and anticipate the sharing of the inheritable traits, it is possible that non-mutants actually mutated themselves by will power, adding a consistent but different emotional mechanism to the hereditary pool of the human-dominated group. Whereas the first mutants would operate by genetic instructions, the second kind of mutants would work out genetically a mode of hyper-excitation of the endocrinal system. This would lend the group an element of obsessive emotionality as soon as genetic miscegenation began.

The social imprinting of shock would come about not by itself alone but in the course of executing symbolic references of the first mutant type, in accepting the obsessive drive of the second mutant type, and in the development of followership among the erstwhile normal band, consisting of sophisticated crowd behavior already possessed by hominids.

All elements would be caught up in the atmospheric reformation. The mutations were consistent with it; they were in fact created by it and responsive to it so that, in a fundamental way, the correspondence of the new world with the new being was assured. Although it did not eradicate the old 'normal beings, ' the radicalized atmosphere punished them and preferred those who responded readily to the new constants.

CHAPTER IV

GESTALT OF CREATION

The human creation happened all at once with a crackling and bursting of the hominidal dam. It was a "gestalt," a configuration of nearly simultaneous and transacting developments emerging from a central change.

A plausible scenario of the birth of mankind might be reconstructed. Let us attempt it.

There follows now a charting of the total process of humanization, to be followed by its discussion.

THE GESTALT OF CREATION AND ITS AFTERMATH
(The Hologenesis of Homo Schizo)

A. Low-powered environmental forces operate, in a uniformitarian way.

B. Hominid is not self-conscious. It has fully functioning instinctual reactions. It has an ape-like cranium, is bipedal, four feet tall, semi-human in appearance, and hairy.

C. Individual concentrates its life energies upon physical well-being and sociability.

D. Perception, cognition and affection are governed by a single coordinated instinctual being. Only rarely and temporarily are they "distorted;" no matter how bizarre or self-destructive its behavior (induced by disease or fright or chemicals) it does not ask "What am I doing?

Postulate now a set of terrorizing natural disasters and distraught faunal populations. Problem now posed is: How could a human be created and survive?

A. High-powered environmental forces are unleashed in sky and earth. All senses are bombarded. Radionic storms change the atmosphere and invade organisms. Physical well-being and sociability are everywhere damaged and threatened. A reign of natural terror..

B. Instinctive behavior is generally frustrated by terror and strange stimuli. Microsecond delays in central nervous system and especially in brain transmissions occur.

C. Schism of consciousness occurs in one or a few hominids with cranial enlargements. Proto-decisions are required for self-control. The "ego" begins as coordinating center for the fragments of the old conscious self; it is actually a poly-ego. Memories are intense. Memories are also suppressed in the struggle for self-control (ego versus alter-egos). Selective recall and forgetting spring into being.

D. The alter-egos displace terror onto other people and the threatening natural forces. The primordial being does not know whether he is "talking to himself" or "talking to others." Self-punishment and self-mutilation are found to be ineffective but persist in efforts "to unite the soul."

E. The ego begins to communicate with its selves by displacement and projection, and having begun the process, extends it to all subjects of displacement. Symbolism as internal language begins. Bilateral asymmetry (righthandedness) is stressed to help centralize dominance in the left-brain hemisphere.

A second phase now occurs, organizing the world schizotypically.

A. High-powered forces continue to impress senses with destruction, chaos, and threats of return. The poly-ego is fraught with ambivalence toward the forces, hence, by retrojection, also among its alter-egos.

B. Perception, cognition, and affection are pliable (less instinctive and internally distorted "thought-disorders") and mix up all kinds of phenomena of the triple-fear (fear of self, fear of others, fear of gods-nature) and triple control system of the person.

C. Principal imprints upon perception of nature and affection are blocking (amnesia, catatonism); compulsive repetitiveness; and orgiasm (destructiveness, wild expressionism). These imprints of the new world order of the schizoid mind operate within the individual, between and among individuals, and between groups and divine (natural) forces.

Without time lapse, a third phase fashions the culture.

A. Persons and groups, so as to control fears of self, others, and the object-world (animated),

B. and to obtain subsistence, affection and the reduction of inner tensions,

C. organize their perceptions, cognitions, affects, and energies,

D. through the mechanisms of memory (amnesia and recall), displacements (associations and ultimately sublimations), compulsive repetition (rites, rituals, habits, rules and routines), orgiasm (aggression and nihilism), and communication (by behavior, signs and symbols),

E. work upon materials and resources of selves, others and the object world,

F. set up all behavior patterns ranging from informal to rigid, including the (1) regime of language, (2) religious rites and structures, (3) compulsory modes of coping with subsistence, sex, and conflict, all of which bear the stamp of the aforesaid needs, fears, and mechanisms but assume variegated culture-forms depending upon the "mix" of history, no matter how brief the history,

G. and then exclude or punish "unaware," "sinful," or "sick" persons or groups who, in relation to a particular culture-mix are deviant (i. e., have too much or too little of the key ingredients),

H. whereupon said deviants (e. g., officially labeled "schizophrenics") must fashion "mixes" of mechanisms and displacements, which, though great in number, represent and resemble in every case the peculiarity of the culture wherein they emerge.

A MIND SPLIT BY MINUTE DELAYS

The gestalt brought forth the prototype human instantly (which explains our use of the world "creation"). Whether by mutation or by trauma, the central event was a splitting of the mind in an essentially schizophrenic reaction. The split mind "recognized its other self." That is, it was forced into a basic, irreversible delusion that it had to deal with an inner person. Self-awareness began. It was an awful feeling.

A permanent blockage (or suppression) was laid down before all instinctual behavior, creating a constant anxiety. The "anxious animal" could no longer act with instinctive ease although it could act more intelligently and with greater versatility.

Now we have the answer to the questions: Why is human instinct so blunted in comparison with primates? How does it happen that all "animal instincts" in humans are within reach of psychosomatism? Instinct is the hair-trigger, set to go off without time for decision-making. Many critical human instincts are reachable by will and can be controlled; indeed they must be. They are set as slow triggers. This happened during the gestalt of creation.

Generalized delays of milliseconds in response time between the limbic and cortical systems and between the left and right brain hemispheres, experienced as environmental, electrical, and chemical impulses, introduce conflicts between the systems and the hemispheres. The delays occur not once, but repeatedly and continuously, because the external forces are not withdrawn immediately. The delays add up to a general depression of instinctive responses, which is sensed by the hominid as both crippling and frightful. Even with microseconds of delay, the organism senses piercing inner contradictions that call for proto-decisions by "itself vs. itself."

A host of proto-decisions fill the behavioral response-space left by the depression of instincts. But there is little experience to help understand, control, and guide the mass of proto-decisions. This new anarchy requires organization, but from what sources and how? A pure anguish, it might be suggested, should drive the hominid back into the archaic limbic system whence no self-awareness would ever emerge. But this cannot occur because the stimuli for the new order of mind have blocked the regression and thrown the bewilderment into the cortical arena.

It is too late to regret the passing of the animal. Either one or another of a pair of cortical referents will triumph by making a decision. Still, the several egos cannot contest indefinitely in a battle of all against all, else, like the warriors who sprang up from the teeth of the dragon that Cadmus slew, they will kill each other. The organism, to survive with its one stomach and conjoined limbs, must act as a whole.

The resolution comes from moving forward, not backwards. The organism widens the gap rather than closes it. What began as a set of millisecond delays becomes an alter ego. The alter ego grows though performance, habit and training into a *weltanschauung* (a world view). The world order emerges, reconstructed by the human mind in a schizoid form.

Drinking and eating, bowel movements, fear-flight-fight, copulation and many other behaviors are animal as well as human, but the human way of performing these operations encases them in schizotypicality. All the behavior that is authentically and ineradicably human is schizotypical.

Impulsiveness begins to become a vice, not a virtue, for the human. The organism comes to realize that at any moment it has the capacity to ask itself questions. As frightful as the experience is, the new human cannot resist the asking. The boundary of the brain hemispheres is the main locus for the sensing of the gap. The left hemisphere, losing slightly its near perfect coordination with the right hemisphere, accomplishes reflection. The reflection is fearful, but effective.

FRIGHT, RECALL, AND AGGRESSION

Fright was all-pervading, both for what was happening inside the person and what happening outside. Because of the terror and the split, a recollective memory leaped into being and with it instant amnesia. Recollective memory was a form of control, occasioned by the delay of instinct. Hominids might remember, but not recall. The voluntariness of recall summoned up the mechanism of the repression of recall.

Meanwhile, the new creature began to talk to himself. (He was, it should be borne in mind, a child without human antecedents). As soon as he thought "what happened to me?" he was human. As soon as he questioned his own behavior, he became superior to all hominids around him. He would think, "I should do this," meaning "we should," and the all-important act of will was born.

Will is the spearhead of the drive to control oneself and the world. Now it was necessary to turn this weapon of will into a weapon of control. For the flood of terror demanded relief. A rapidly growing stream of symbols crossed the bridges between the two selves and flowed out to attach the symbols to the outer world and especially that part of the outer world that was threatening destruction, the turbulent skies and the effects they were producing on Earth.

Great fear was never to be eliminated from the human. It dominated his mind and set limits upon all of his behavior. It was the fear of his own schizoid character and fear of the outside world (and the gods). Of all unpleasantness, being two people is perhaps the most continuously unpleasant. Out of such fear comes the desire to control and somehow stabilize the situation, preferably by merging dissimilar selves into the original unity. Assuming some success in achieving stability, any increase in internal or external fears will excite the fear of loss of self-control.

In all of this a large role for human aggressiveness is prepared, for the world must be controlled if anxiety is to be relieved. Or one must delude oneself into believing that it is controlled and that one can take part in the control system to insure that it will work. It was a formidable assignment. Still, for a madman nothing is impossible, as we shall see.

When the sensitive brute could not endure the intensity and scale of internal and external disasters that confronted him, he explored, besides flight and fight, other means of control to cope with "reality." And immediately upon seeking control he found it in the new exigencies of his constitution: in the ability to recall and forget, to perceive his individuality and duality, in "flights of fancy" and in the symbolization of his lines of communication within himself and between himself and the outer world.

When finally given respite from panic, these mechanisms could be used pragmatically, with brilliant and instant success, to organize and invent for all aspects of life. The human had become unconquerable, and lusted for conquest.

Ordinary animal fears, with which hominid was not unusually beset, given his many abilities, were inadequate to move him into a new phase of development. With its uncontrolled and widespread displacement, the great fear, however, threatened all existence and, by inference, every life-value of the organism - procreation (sex), health, food, sense of control and adaptability, and affectional ties of the primal horde. Hence, the changed character of the mutant human affected all life-values and thereupon all the new institutions that came to be.

The very fact that the changed hominid could reflect upon itself meant that it was not itself, but split self. So to primordial fear was added existential fear, the fear of one's own self-awareness, the distress of standing off from oneself, the basic schizophrenia of humankind, largely delusory from the standpoint of physiology since the same organs served the plural selves, but of course the schizophrenia was itself physiologically founded.

The origins of human nature were connected with the fearing components of hominid nature, and the subsequent history of human nature, as a result, has been mostly unhappy. The misery is generic, and therefore persists even when the rude clutch of disaster is released, as it was for periods of time, early and lately. The structure of the readily mutable mammal, the hominid, was such that a "benevolent" mutation, if conceivable, might have been utterly destructive. Generally, nature adds in evolution; it complicates; the easiest thing to capacitate in man was his brain; so he got a multiple head.

"What happened to me?" was the first question. Then came the gestalt of creation: it was composed of awareness, symbolizing, and projection. The proto-human strove to recollect himself amidst the turmoil of his kind and of nature. To exist and survive he had to discover himself amidst the disruptions of memory. His subconscious now existed in a way that it had not before, as a well of confusion, that overflowed with images that did not belong to the present, that offered uncoordinated seemingly unrelated elements that were taken care of by "unmechanistic ways" unfamiliar to animals. Before he could say "I am," or "I think, therefore I am," he had to come to terms with a new subconscious that distorted all perceptions of himself and others. His character was born of delusion.

The broken mind of the beast sought to restore itself, but could only do so under new terms. Restoration of the previous state was destructive to the organism. A consciousness had to be organized to seek materials to guide the organism in its disorganized condition. It had to pull what it required from the forgotten, which was not really forgotten, but which would no longer normally emerge in a flow of instinctive, directed, utilizable unconscious information that characterized the hominid. The tortures and triumphs of memory then began. The accommodation of an awareness to an uncontrollable but recognized history began. The conscious and the willful assembled together upon this small island in a sea of suppression.

Dominating the transition from a brutish to a human character was the psychological mechanism of projection which sprang from the creative gestalt. Projection is the imputation to another perceived existence or being of one's own motives and wishes. Once projection is achieved, and a world of transactions, real and imaginary, is set up.

As his own self divided through self-awareness, man's gaze was fascinated by the sky. As he questioned himself, he questioned the intangible and uncontrollable world of the skies and all its mundane effects. The fact that the heavenly forces were abstract and impersonal became a matter of concern much later for a few generations of philosophers. Primeval man did not own a neuter gender. Everything in the world was alive. He did not have to acquire an illusion permitting him to reify or anthropomorphize. For he never had made and had now no reason to make a distinction between the living and the inorganic.

Projection to objects as living things was immediate. The gods came into being. What traits the gods came then to possess were the actual traits of a god as witnessed, the traits (later on) of remembered gods, the feelings and traits of mankind in chaos and birth, and such traits of life forms on earth as mankind perceived and found to be analogous to his own and those of the gods. What he saw in the sky confirmed and strengthened his projections and let them be retrojected into his own traits even more strongly. Each time this happened, there was a self-fulfilling prophecy, a growing obsessiveness, an enhanced belief that one was being threatened by sinister forces (paranoia).

SELF-CONSCIOUSNESS

Self-consciousness, the poly-ego, was a village built upon piles driven into the sands of permanent existential anxiety. It was and is a patterned and integrated architecture accommodating to the neural blockages that deter instinctive solutions. The neural blockages are stabilized by socially elaborated mechanisms that take certain forms such as rituals, theology, and logic. Further, self-awareness involved the use of symbols, first to institute an inner communication system, and then to introduce transactions with others and the outer world so as to keep the far-flung egos fully operative. Once achieved and begun, physiochemically constituted and socially founded, self-consciousness is revived post-natally in each generation.

The human poly-ego was both individual and social. Those possessed of it sensed themselves unique, and at the same time identical with their groups. It produced an anarchism at one extreme and a regimented discipline at the other that go far beyond the capabilities of the mammals. Herein lies the eternal cooperation and conflict between the individual and the group, which is the subject of so much philosophy, sociology and political science.

Self-consciousness in humans is not only awareness, as in a wakeful animal, but it is awareness of the (other) self or selves as entities. Further, the self looks at itself and at other people and objects with the same dynamic. Thus, in terms of psychopathology, the self -- the poly-selves -- is a form of delusional thought in the schizophrenic category of the split self. This usage may have developed as a convenience for considering therapy; but in actuality the poly-ego is the only human self to exist and is a system of normal and sane delusions. Since self-consciousness did not exist until the catastrophes began, the fundamental breakdown occurred only once, this in the days of creation. Repeatedly, in subsequent catastrophes, the mind might drift from its first moorings, but, with the help of culture, it would arrive at another anchorage in a new set of self-conscious delusions.

An ancient set of events is incorporated in the story of Adam (man) and Eve (woman); after having eaten of the tree of knowledge, they became shameful in their own eyes, shameful for their nakedness. From a blissful lull of unthreatened self-consciousness they passed, under the harsh command of their god, into a renewal of their self-awareness. In the millennia before the new disaster struck them, they had apparently developed a religious and symbolic world of a humanistic kind. This was the Garden of Eden, a "Golden Age" to other cultures, where apparently threats to the poly-ego, now stabilized, were few. Analogously, Giambattista Vico comments that it was the thunderbolting electricity of Jupiter that produced the first Muse, who defined "knowledge of good and evil," a power only later called "divination,"[127] which then, much later, is regarded as a kind of superstition.

In the age of Yahweh, perhaps millennia afterwards, new catastrophes of Exodus and the wilderness occurred, and the Hebrew Deuteronomy declared, (28.27-9), "The Lord will smite you with madness and blindness and confusion of mind; and you shall grope at noonday, as the blind grope in darkness." Thus whole groups of people might lose their ordinary minds, but never their human minds. Typically, the blows bludgeoned the self-aware mind into extreme pathological states (in human terms), but afterwards the mode of recovery was always the same.

The mechanisms of the human conscious proved to be functional not only in obtaining relief from anxiety, but also at the same time in providing the goods of life. The proto-typical "madness" was "superior" for coping. So the mechanisms of the conscious found themselves to be generally released from their total service to emergency needs of disastrous times. They came to be used pragmatically for many other purposes, including the development of the useful arts and crafts and for social organization. Ultimately there occurred an everyday dissociation of the emergency and pragmatic functions of the self-aware ego.

The emergency functions, that are similar in effects to the superego, are more particularly the primordial functions, ordinarily engrossed by theology. When a new disaster occurred, when the polyego system again was deeply disturbed and dissolving, the old self displaced the new pragmatic self and recapitulated the mechanisms of defense as they were employed in the days of creation. In times of great stress and fright, it is the primordial human self that takes command, not the unconscious nor the beast. A human organism will fall into a catatonic coma or die before releasing the self-consciousness it received upon creation. It will temporalize, symbolize, and control, up to the brink of eternity. Oblivion marks the surrender or death of self-awareness.

---

[127] *The New Science*, 82.

The problems typical of the human species are in the regression of the ego-mechanisms to their primeval but human state, and not in the resurgent total triumph of the hominid. The unconscious, when reviewed, is seen to be the reservoir of hominid instincts and the suppressed or forgotten materials of experience. However, it is commanded and transformed by the primeval ego, even though physiologically coordinated with the aboriginal instinctive animal.

MEMORY AND FORGETTING

Whence might come such lines as these of Baudelaire?

> In those days when Nature in her bursting vigor
> Bore of herself each day such monstrous children
> I would have loved to live with a young giantess
> As at the feet of a queen, a voluptuous cat.[128]

The corridors of art and culture everywhere echo with the cries and gasps of remote recollections.

R. G. Hoskins, in his essays on schizophrenia, writes how patients describe their mental illness:

> Very commonly it is as if the conscious self had descended to some lower region where it is no longer in control... The eyes are opened so that one seems to see back to the beginning of creation. One seems to have lived perhaps in many previous existences.

Mnemosyne (Memoria), according to a Greek legend, was the daughter of Ouranos but she bore the muses of the arts and sciences from Zeus, grandson of Ouranos and a much younger god. Thus Greek cosmogony assigned memory as an immediate effect of creation. Memory would have begun in the self-awareness of the gestalt of creation. Heavy terror worked to forge memory and forgetting. Out of the material of all things, it hammered the deepest memories.

---

[128] "La Géante," in *Baudelaire*, Scarte, ed., Baltimore: Penguin, 1961, 25.

There was too much that was too bad to remember easily, and it was forgotten.[129] Also too much was forgotten for even the unconscious sectors of the mind to bear. Recall may be regarded as the most obvious and 'rational' function of the memory mechanism, mnemotechnology; it operates, however, only if the recollection does not destabilize the poly-ego. The 'forgotten, ' that is, the memories that tend to destabilize the ego's confederational balance, provide essential subject-matter and forms to sublimate activity. They force their way into remembrance via the routes of theology, myth, literature, the practical arts and sciences, and social behavior.

The pride of man in a memory that is superior to that of the beasts is inordinate. Memory is a weak, self-imposed tool for displaying material to the conscious in a light that poorly reveals its sources. The special human memory, like everything else uniquely human, is a device that the beast may not need. But the human must have it. He does not go around picking his fundamental qualities like pretty spring flowers in a meadow.

Accompanying the primary amnesia of events themselves, is a secondary amnesia that is associated with the forgetting of events. The amnesia of man came from the primal terrors and set up the mechanism of denial, which first insists that nothing was forgotten, and then persists in denying all sorts of traumatic memories, until we find him today the congenital liar, lying both consciously and unconsciously.

Man does not remember his experiences as Hominid 'X, ' because the hominid had a conditioned reflex system that typically registered reaction, not selection; they were too boring and useless to recall; they were not layered by meanings, symbolized or acculturated; they were not history.

Further, the shock of humanization was also a shock of de-hominization. Forgetting that we were once hominids is part of the amnesia of the trauma of creation. The autonomous system of selective (though usually only apparently so) memory began with the creation condition which we chose to remember and the sublimation of the larger part of the events. All cultures have creation stories. Before creation, man was clay or animal or part-god. The gods, they tell us - and what gods are not crazy - give us our special schizoid minds.

The memories were in the brain, memories of all except the most trivial and fleeting of events. They were diffused around the brain but could not be called up indiscriminately. The call had to come from one of the poly-selves and then would be subject to a veto from another self or from the central government of the selves. An endless complex track or network

---

[129] A. De Grazia, "Palaeoetiology of Fear and Memory," in Milton, *op. cit.*, 31-46.

became probably an index of symbols, an inner language. When this language was developed as a political process, in communications with others, it formed an outer or public symbolism or language. But this is only a small fraction of the inner language that connects memories.

In the public language that ultimately developed were contained clusters of words that grew into creation stories, which purport to describe the days of creation of the world and of humanity. We expect the stories to be heavily veiled accounts of a true history, much like dreams that are internally distorted and censored but nevertheless lend themselves to scientific interpretation up to a degree. We see in the need for creator gods a determination to tell the truth in some way, to assert that the human was distinctly born.

Also stories were told of the environment before and during creation. In the new public language, the legends agree that there was a chaos, a formless, mindless order of the world, going nowhere until the divine intervention. We see two types of important 'fact' in this chaos. First, there is a reality, an Earth with a dense translucent encircling high firmament clouds. Second, the human is not there but is about to appear. Man appears as the canopy breaks and the gods appear.

Beyond this core of agreement, which I have not fully described here, the stories diverge. Running them together is like reciting a stream of dreams, all apparently referring to a single theme. This earliest extant public language is just what we would expect it to be, and what dreams are like, too, and what the world often appears like to persons suffering from mental illness. They hold a truth which can be deciphered.

The modern schizotype or schizophrene may get up from bed late because it takes hours to sort out his dreams from his reality. Primordial homo schizo must have had the same problem, and, if it were not for the fact that primates waste a lot of time anyhow, the new human might have been victimized for this trait. But, on the 'positive' side, he acquired many new displacements (by analogy) from his dreams, a great many wish-fulfillments that encouraged his ambitions to control the world, and a number of incredible (to us) believable orders (to himself) to sally forth and conquer the world.

To dream is to sleep, and, as the poet says, to sleep is to dream. Tinbergen says more about sleep in humans:

> Another innate displacement activity in man seems to be sleep. In low intensities, in the form of yawning, it is of common occurrence in mild conflict situations. Just as in some birds (avocet, oystercatcher, and other waders) actual sleep is an outlet in situations where the aggressive instinct and the instinct of

escape are simultaneously aroused. Reliable and trained observers, among them Professor P. Palmgren of Helsingfors, have told me that in situations of extreme tensions at the front, just before an actual attack, infantrymen may be overcome by a nearly insurmountable inclination to go to sleep.

I can attest to this, having sleepily observed sleepers under the circumstances. I note, too, that the Spartan warriors at Thermopylae, having sent home their allies and decided to die in the approaching overwhelming assaults of the Persians, spent their time dreamily, mutually combing their long tresses, much to the surprise of Persian scouts. Tinbergen adds, then, that

> Sleep, as is known from Hess's experiment, is a true instinctive act, depending on stimulation of a centre in the hypothalamus. It is also in line with other instinctive acts in that it is the goal of a special kind of appetitive behavior.

The human is sleepier, or at least sleeps much more deeply and determinedly, than animals except when these hibernate. Sleep, dreams, hibernation, self-hypnosis in crisis expectancy, drifting hazes of schizophrenic displacement, catatonic cultures, sleeping culture pockets, and retreat from the dreaded or impossible; there is here an interrelated complex that helps to index some of the catatonic control operations essential to homo schizo.

THE STRUGGLE OF THE SELVES

The ancestor of homo schizo carried a bilateralized brain; two generalized and equally functioning hemispheres operated with a minimum of conflict. The cerebrum of Hominid 'X' was large, perhaps too large already to escape conflict arising out of intra-brain and central nervous system dyscoordination. Homo schizo inherited a larger brain, with immediate problems of electro- chemical and nutritional supply, and of neuro-transmission speeds. In homo schizo the brain conflict evades the earlier physiological compensation by moving out in all directions. So it gets less rest, is more continuously restless, awake and asleep. Rest often takes the form of diversion. As already pictured, specialization within the brain was sharply increased, and right-handedness developed. A general feeling of fear, inadequacy, and weakness was instituted that demanded obsessive attempts at self-control, which extended outwards as attempts to control the environment.

The mind of the hominid was shattered. The quantitative leap was great enough to be termed a quantavolution, a qualitative change. One of the sometimes enviable and endearing traits of higher mammals is their consistency of behavior. But we are often so bored with this quality that we search for the smallest indications of character in horses, dogs, and apes. After a night in town a drunk can mount a donkey and be carried asleep over the mountains to home; the animal is 'given his head.' Many species, we note admiringly, have 'minds of their own.' Actually, they do not; it is easy to trace the instinctive sources of the behavior. They do not perform many behaviors where doubt and decision are present. We can assume that the hominid was not plagued by indecision nor driven by strong needs to control himself and the world.

The human mind that eventuates is a troubled regime. The ego is a would-be dictator whose position is shaky. He can be toppled at any time when his foreign possessions - the outer world - revolt and attack him, and his inner subordinates have sufficient autonomy to join the foreign alliance or to launch a rebellion on their own initiative. Hence we say that the hominid mind broke down in quantavolution and the human ego, basing itself on the large 'lower-level' elements of the central nervous system, grew out of the chaos of the "higher level" elements. The ego, then, was never hominidal and never absolute. It came into existence as a suzerain and would-be dictator, and can be toppled or changed as its components grant or withhold loyalty.

A great step of the suzerain ego is to consolidate its control by seizing and managing the right hand. In a typical neglect of transactional philosophy, it is conceived that righthandedness is "logical" inasmuch as the right side of the body is controlled by the "dominant" left brain hemisphere. May it not be more logical to conceive of the right hand as being developed by the left brain in order to strengthen the dominance of the left brain? Right-handedness is genetically predisposed, but only because the left cerebral hemisphere is genetically dominant. The left brain commissioned the right hand to be commanding officer in order to bolster its shaky regime.

Similarly, a struggle of the selves took place outside the mind, in the environment, in outer space. In this arena, we see the stars and planets, the comets, the clouds, the moon and sun. Homo schizo first saw these objects in a way that no hominid could see them. He was, we recall, striving to establish a dictator-ego, preferably to carry himself back to his golden age of instinctive bliss. The situation was, however, chaotic, and other selves were offering themselves as candidates for authority, or worse, were practicing anarchists.

Here is where symbolism might play a major role as an ally of the dictator ego. If everything was to be called by name, and the code for the names were locked in the code counting and sorting computer of the brain, then whoever held the computer key was the master of the brain and body. So language was seized upon and developed by the left brain. With symbols and a strong right hand, a viable regime could be and was established.

Too, there were no limits to the symbolism. As fast as fear erupted and displaced itself, even to far space, the symbols could pursue it and control it. To name an object is to rule it. Always the principal ego was to be an uncertain despot, yet to be a magnificent one, on whose infinite territory not the sun, not even the stars, ever set. The substance of all of these operations may have taken time to occur and be realized by the self-aware human. But the implications of them were already present upon the gestalt of creation.

BECOMING TWO-LEGGED

Humans probably became totally committed to stand and walk on two legs upon genesis. They were shifting their anatomy to conform to the global reconstitution of their mentation. Students have now shown that australopithecus was bipedal, and feel confident that homo erectus was as well. Hominid 'X', the common ancestral form of them and the human, can be imagined as preadapted to the point where he might, if he would, be bipedal. Like handedness, as soon as the ego-struggle occurred, bipedalism was pressed into service by the dominating left brain.

The human stance is unique, but the anatomy of standing is only presumed to be unique. We remind ourselves that the Indian feral child, Kamala, was totally adapted to quadruped motion to the age of perhaps eight years, and several years of training were required to get her to stand and walk voluntarily. Her muscles, tendons, hands, feet, knees, and probably her total body posture were quadrupedal. She walked on her tough palms with a full heel-to-toe motion.

A number of physiological and anatomical changes accompany bipedalism, but perhaps all are ex post facto, such as the stretching of lungs and swelling of blood vessels to the head. Most likely, bipedalism is an adaptation for which an intense determination is required. There are no commonsense reasons for it. Kamala was comfortable on all fours and could run well. The human infant, of course, crawls for a year and more before being able to stand up and toddle. Only for sophisticated human activities is bipedalism superior, which presumes that humanism came first. Primates and other mammals are physically and socially more intimate than humans, even including the great cats within their own families; they might be called more 'tangiphile. '

Bipedalism had some motive in the schizoid complex, in which aversiveness to others and ambivalence are prominent. Standing erect is a gesture of retreat and removal from others, which individuates beings. It is also a threatening and offensive posture, including the conspicuous chest-thumping that fiction-writers overrate in gorillas. It goes along with (deliberately) smelling less, and with offering less in the way of hindquarters and front-features to nuzzle and smell. It encourages genital privacy because the hand and upper torso can exercise protective movements. The first homo schizo, one may conclude, would voluntarily seize upon bipedalism, if it were not an already confirmed behavior.

Bipedalism, therefore, matched the character of homo schizo and he is determined to master it. But what was this determination or voluntariness or will? Man was supposed to possess a will; philosophers and *hoi polloi* have thought so for thousands of years. Recently, however, the will has been removed by the philosophers of determinism, although retained by the masses. Did events occur during the gestalt of creation to give humans a will, yet permit it to be taken away under later rational analysis?

VOLUNTARISM

The 'will' in hominid, we postulate, must be a 'want, ' typical of animals, therefore an instinct - basically a will to feed, fornicate, flee or fight. In the disaster of creation, the new human achieved a new primary 'want, ' to control himself and the world, to rid himself of fright. All of hominid's will - the aforesaid 'Four F's' - is subordinated, rendered secondary, to the primary will to control.

Since the will to control is conveyed to a bewildering variety of human displacements and identifications, it acquires a new complex aesthetics that deludes humans as to its nature. People (philosophers and theologians among them) came to think that they were dealing with a qualitatively distinct mechanism, whereas it was a highly diffused aspect of all human activity, capable of exponentially more fixations than the simple 'Four F's' of the beast. Some acts of obsession and compulsion came to be called 'will, ' when they pertained to objectives of positive or negative value. These, however, if we ignore value preferences and their large variety, can be reduced to the great gestalt of instinct-delay, split self, existential fear, and consequent promiscuous and obsessive need to control.

The world is as will, then, just as Hegel said. It is a delusional creation of man's poly-ego confederation playing with its kaleidoscope. This game, with its dexterity and intensity, put all other animals to shame. And individual men came to be distinguished infinitely, in their applications of will, by the way their particular minds shook their kaleidoscopes. So that one man's iron will was to win a battle, another's to win a certain mate, another's to gather money, another's to die, another's to conquer will itself by willing nothingness. Much of this diversity probably occurred promptly after the time of the primeval gestalt. Its diversity elaborated into virtuosity, which doubtlessly played a part in intimidating all surrounding conscious animal forms, including our erstwhile hominid cousins.

DIFFUSION OF THE GESTALT

The hypothesis pursued here is that the gestalt of creation happened to one or two hominids, and diffused as a new dominant gene system. Were this proven untrue, we should proceed to a hypothesis of changed atmospheric constants. If this should be proven incorrect, we should retreat to a theory of psychosomatism, that combines psychosomatism, the 'omnipotence of thought, ' and potentiation of everpresent lines of development of essential living matter. If this idea should be overthrown, we would put up a last-ditch defense with a purely cultural theory of catastrophic fright overturning the hominid mind. All of these are conceived to have been quantavolutionary changes, occurring quickly and hologenetically, from the one Hominid 'X' species to the present homo sapiens schizotypus. Further, it is likely that elements of all of three entered into the actual rise of homo schizo and his further development up to the gates of history. The theory of mutation-by-mutation, adaptation, rung-by-rung, millions-of-years' evolution that is generally held today seems to be mistaken and useless.

In the quantavolution of homo schizo, what happened to the Hominid 'X' ancestors, and to diverging strains such as homo erectus and Neanderthal? Many mutated, sickened, and died under the catastrophic conditions that were required to generate the new dominant gene system of mankind. Furthermore, the character of the new species was such as to intimidate the hominids and drive them into marginal living niches. Inasmuch as interbreeding was common, the human population would contain for some centuries or millennia hominid members and human members with hominid genes.

The sharp differences between the two types of creatures would encourage eugenics as a matter of course. There are many examples, in social and historical practice, of obligatory or authorized infanticide and of celibacy enforced upon special groups, tribes, serfs or slaves. Holy wars have been many. The hominids, then, insofar as they were not eliminated by segregation and extirpation, could have been subjected to absolute interdiction by the rules of birth and social nurture.

In a quantavolution by atmospheric change, the scenario of the gestalt would have been replicated in many hominidal settings. A number of humans would have promptly appeared. The transition from hominid to *homo* would nevertheless proceed under the conditions just stated. Might the mutations required for humanization have occurred in several hominidal settings, and thereupon and later be fed into the human gene pool via miscegenation? We would then witness, for example, a fire-making band joining a speaking band, from which speaking fire-makers would be born. But the theory of homo schizo requires that his traits should fall out from a central trait change, which we have pinpointed as the splitting of the self. The single genetic incident is fully explanatory, and it cannot admit of any but minor exceptions to the hologenesis of traits.

THE DOUBLE CATASTROPHE

The necessity of natural catastrophe has been recognized, if a critical mutation of species is to be experienced. In other works and in earlier pages, I have presented the theory and evidence for such catastrophes. Strictly speaking, this external catastrophism is distinct from the internal catastrophism of creation. Man is a catastrophized animal: both external catastrophes and the internal catastrophe of his genesis have awarded him this title.

Confusion between the two types of catastrophe can occur, as it did in some earlier passages that I have published. For, not only is there the outer chaos and the inner chaos but there is also the overlapping of the natural catastrophes with the earliest experience of homo schizo. He speaks the language of catastrophe out of experience.

Critics of quantavolutionary theory can turn this around and say that homo schizo, being what I have said him to be, naturally imagines all kinds of natural catastrophes to have occurred to which he was witness; that is, he would normally have hallucinated world-destroying catastrophes. For instance, Fenichel alludes in his *Psychoanalysis* to the manic's desire to control the world and Sebastian de Grazia in his *Political Community* to the ever-present ideology of the destruction and reconstruction of the world. Can I not keep the skies swept clean and in order, leaving catastrophes to occur solely in the mind?

To this, I would respond that homo schizo's stories of great disasters are too well supported, and too well detailed, to be either imaginary or highly exaggerated tales. It might be expected, too, that people who were genetically frightened, to appease their fears, would tell stories of a golden age and a gradual progressivism of mankind, which they do; these are partly there, but by their temporary historical framing they lend support to the disaster stories, so that both types of recollections must be accorded historicity, and thereupon further analyzed.

Because the terrors were sensible manifestations of high-energy forces, delusion and reality were forever commingled in the new species. The range of thought and sense material was great, including as it did the practically infinite combinations of sense data of the high-energy events and the immediately and infinitely symbolized associations of the events with the self and group. Not only are the earliest records loaded with catastrophic events and languages, but so also are Shakespeare's comedies and tragedies.

The great variety of detail in man's innumerable culture traits is an expectable and understandable resultant of all the psychological and real events attending the creation. For every controlled and uncontrolled construction that the mind emplaces upon events and objects, there are real events to fit into it. De Santillana and H. von Dechend explicitly commend a large, though unmeasurable, quantity of historicity in Ovid's work on metamorphoses. The palaetiology of the concept of metamorphosis may rest upon an abundance of mutated and damaged organisms accompanying atmospheric and radionic disasters. To hear it told, life was never dull *illo tempore*.

A portion of all religious expression and practice relates to such quantavolutions, among other things. These *spectra horribilem* then serve as religious lessons, teaching groups and individuals of the punishing power of the gods. The same events serve to connect the celestial with the mundane, inasmuch as sky images and stars are connected with the mostly terrible changes. The lack of control over mutants raises the level of terror. Therefore, what appear to non-quantavolutionists to be unconnected, inconsistent, and unexplainable varieties of the production of human minds here and there throughout the world, acquire under quantavolutionary theory a simple logic within a single framework of explanation.

A PRIMORDIAL SCENARIO

Ideally, the general scenario of the hologenesis of homo schizo would provide a highly specific scenario such as the following:

A pregnant twenty-one year old female of the species homo erectus *frater* (that is, Hominid 'X') is a member of a band of thirteen that gains its livelihood by gathering nuts, berries and herbs, and hunting small animals and choice insects in a swamp habitat. It eats roasted products of wild fires, even spreading them to harvest a territory. It has no tools, not even reusable clubs. An aggressive older male leads the group, which has hegemony over some fifty square miles of territory. The group straggles about. The large mammals hardly disturb them, for they put on a brave front, screeching, gesticulating, baring their small teeth, and dodging adeptly. They are tree and rock climbers, and swamp floppers. They are, in effect, too much of a nuisance to bother with and not tasty to eat.

But as the camera zooms in upon an abri, laying off a swamp, one female, 'Ma, ' is dropping an infant. For a long time, which no memory exists to appreciate, "the skies have been falling;" the waters are rising; fires are frequent; volcanoes are bursting asunder; and the animals, as always now, are agitated. They do not know it, but the fall-out of radiation is heavy. Many die without seeming cause. Many infants are born dead. Many dead animals of the water, sky and air are discoverable and eatable, though some may be radiated and chemically toxic. So living is easy, but stresses are heavy.

Ma bears forth two monsters, identical twin males, glabrous, their heads noticeably larger, their movements and cries unprecedented in volume and queer. Ma and others nurture them and they survive.

They are the bane of the band. They seem never to grow and their demands are insistent and unending. Cowed by them, the band cannot kill or abandon them, but as if by order of a superior, give them what they ask for, so far as possible. They are tough and wiry but not a match at first for the other young of the band, who begin to breed before these are mature. Still they have a strange power and dominate most of the band, exhibiting an aggressive acquisitiveness. Their command of screeches and gestures is far superior to everyone else's.

They behave in unexpected ways. They will carry fire farther, preserve clubs, go out of their way to spy on game, remember the nature and sources of comestibles, pack, store and carry provisions, and use their resources aggressively to dominate the whole group including the present leader. They hurl pebbles at friend and foe alike. As if they can see how they appear, they stand on their hind legs and howl needlessly, with their right arms shaking a club, apparently with intent, at the sky, at holes in the ground, and against the winds. They do not forget, and discriminate savagely among the group, for one thing raising Ma to a status higher than that of anyone else, male or female, rewarding her for favors long past.

The time of reproduction comes, tardily. The siblings, who have fought off together the approaches of others, mate with their mother and every female about, and other monsters come forth, bawling. Nothing is too good for the several mutant brats that issue, and their pressure for variegated responses is such that the band actually loses hominid members, by premature death and fighting. But in the next dozen years, the band grows by ten monsters, several of them out of hominids by the male mutants, and includes only six servile hominids, who are treated like retarded children.

The mutants prattle incessantly among themselves, gather and hunt successfully, carry flags or branches that intimidate even large animals, not to mention other hominids, give Ma a decent and jealous burial, then dig up her skull and set it nicely in a niche of an abri that has become their headquarters, surrounding the whole with rocks and letting only the docile enter. When the group goes off on long journeys, the young, the sick and the old are left at the abri, comforted by Ma's skull and continuous fire. They spend their time attracting living things to their garbage pit and dispatching them; they chant, they make rope, break stone, and whittle lances.

There is little more to be witnessed. As we take our leave, we are satisfied that a human culture, up to the standards of the twentieth century in most respects, will manifest itself in a scenario of fifty years into the future.

The mutants - call them homo schizo - ill number three hundred, living among a dozen bands, ruling these and drawing the remaining hominids for services, and tribute, possibly cannibalizing them when convenient. The infectious family will have seeded the most attractive of the females and spread out for a thousand square miles around.

Some hominids who are docile, or children of the mutants, remain; their germ plasm will soon carry schizoid genes and they are themselves trained to resemble the homo schizo types in behavior. The others flee or are killed for resisting progress in some sphere of life. Unlucky the hominid band that broke away with no mutant.

Large animals can now be trapped and killed. No natural enemies exist, except microorganisms, to threaten survival. There is, however, the enemy within, for homo schizo, when seized by the will, attacks his own kind. Several bands are to be found hundreds of miles away, led by people who have fled or been driven from the homeland.

Only the gods above who animate the violent forces of nature are respected and communicated with by declamations, exclamations, obeisance, gifts, chants, and dancing. This polymorphously perverse people, their instincts unleashed, are driven to try whatever comes to mind; they are capable of stressing themselves inordinately and setting up and breaking down habits continually.

If the reader is interested in comparing scenarios, he may refer back to the "evolutionary ladder scenario" set forward earlier or to one of the "quantum speciation" school of thought, in Steven Stanley's *New Evolutionary Timetable* (157-8).

QUANTAVOLUTION AND HOLOGENESIS

The human probably was born from Hominid 'X' in a brief incident that, for reasons given elsewhere, might be placed at 13,000 years ago, perhaps even a millennium or two later, but also perhaps within a 300,000 years period earlier. It would be well to fix the Holocene boundary at the point where the humans appeared. The aggregate of data on australopithecus and homo erectus promotes them to adjunct humans, also descended form Hominid 'X. ' Hence, in the preceding chapter, they were projected up the ladder of time.

John Pfeiffer, in some unusual passages, tells of how competent are the economics and how full the minds of the people of today, the Bushmen and the aboriginal Australians, deemed the simplest of humans, though living in an environment incomparably more difficult than what it once was.[130] He reports on their high mobility, the thousands of square kilometers over which they regularly range.

He tells us too, of the charming dream of Louis Leakey, of a kind of dynamuseum, as I once termed such, where visitors would, each week, be transported into a different stage of human development, living as an early australopithecine one week, and the next week according to another way of life. Week by week it goes -- as if time could be collapsed and we might develop so quickly, which is true enough to be suggestive.

The homeland of mankind cannot yet be ascertained, even though we agree with Washburn and Moore that man was born only once, at one time, in one place.[131] We speculate that out of Hominid 'X', whose behavior and appearance were distinctly different from those of the "hominids," "proto-humans," and modern humans of whom we know at present, there came a macroevoluted or quantavoluted type who intermingled with and dominated these families in short order. In *Chaos and Creation* I drew a schematic diagram of the continents of the Earth as they were once gathered

---

[130] *Op. cit.*, 210.

[131] *From Ape to Man; cf* H. H. Wilder, *Pedigree of the Human Race*, N. Y.: Holt, 1927, 156-7; E. A. Hooton, *Apes, Men, and Morons*, London, 1938, 185.

together in an all-land world. In this Pangea there occurs a location which can only be imagined today because of the ocean's opening up and the continents separating. The Caribbean region and the entrance to the Mediterranean dividing Europe and Africa were probably a single landed area with shallow seas, the legendary and geological Tethyan Sea.

This kind of area can be regarded as a possible original home of mankind, and I shall sketch here an idea of it. Basic to the argument is that Hominid 'X' existed in numbers everywhere and became human before the globe cracked, before the continents moved to their present positions, and that all of these events happened between 14,000 and 11,000 years ago. The defense of this time scale is carried in *Chaos and Creation, The Lately Tortured Earth,* and *Solaria Binaria.*

Our choice of location may be preferable to the African rift, a treasury of early finds because it has been exposed by geological erosion. Our guess may also be preferable to the speculation of Thomas Huxley (accepted by the polymath co- founder of communism, Frederick Engels) that mankind originated in a now sunken area called Lemuria, a presumed tropical zone of the West Indian Ocean alluded to in Indian and African legend; this idea does not account so well for northwestern man. The location is more likely, too, than the high Iranian plateau, which more plausibly provided a refuge for disaster survivors and only much later a mobilization area for the later descent of Indo-Europeans towards the west and south.

A race 'Atlanticus' may be represented in the proto-Mediterranean type and the aboriginal Europeans, North Africans, and seemingly Caucasoid traces of types reported in earliest American depictions and myths. The homeland is postulated at a point not too far from the focus of Atlantean legends. It follows educated guesses by early anthropologists such as Frobenius, who thought that man moved first from West to East and then back in later times. Nor does it contradict the evidence of relative movements and superposition of fossil data in the fossil and cultural discoveries of the past fifty years.

The Americas are usually considered to have been barren of human life until Holocene times, or until late in human development. I think it more likely that existing incidental evidence of man's presence in the Americas will ultimately be augmented to the point of acceptance. At present we have hundreds of items such as inter-racial picture albums (Wuthenau's Unexpected Faces), an incredible upper left second molar associated with pliohippus and other Pliocene animals, in Nebraska (evaluated between pithecanthropus and Neanderthal), and Hooton's claim of finding negroid skulls among pre-Columbian inhabitants of Mexico.

With a compacting of time, what appear to be long gaps in human development will disappear as illusions. It is probably no more implausible than other theories, that australopithecus, evolving with its Hominid 'X' form, and neo-humans moved through the then tip of South America, also down throughout Africa, thence through then-joined India, Madagascar, Antarctica, Australia, and eastwards, also, through what is now the Near East, Arabia and the South Asian islands.

Neanderthal's mixed hominidal-human group would have moved eastwards following the shores of the Tethyan belt through Turkey, Iran and China. Homo erectus, in combined human-hominidal form, would have struck North and South to the farthest extremities of Greenland and South America, and in a wide sweep westwards through Africa into the now South Asian Islands and farther north to China and beyond.

"Beyond" here means, by the Pangean theory, all the land, into which elements of all races found their way, which was exploded and blasted away in the greatest of catastrophes, that which saw the material constituting the Moon pulled out largely from what is now the Pacific Ocean Basin. Once again, the reader is referred to the statement of this theory in the aforesaid volumes.

I have mentioned earlier the controversial works of Ameghino that claimed an extremely old date for the fossils of men of the Pampas. Nor can the halt always lead the blind; a radiometric dating by the uranium-thorium method gave an age of 81,000 years for a human tool made of mammoth bone. It is from Old Crow Basin in the Yukon, and was reported by Richard Morlan of the Ottawa National Museum of Man. In California and Mexico, claims of around a quarter of a million years of age have been made for two sites of human operations.[132] Artifacts at the Calico site (California) were assigned by uranium-thorium tests an age of 200,000 ° 20,000 years. Similar dates were assigned by both fission-track dating of volcanic material and uranium dating of a camel pelvis to the Hueyatlaco (Vasequillo, Mexico) site containing sophisticated stone tools, by a second group of scientists.[133] Once again radiometric dating is thrown open to question, but also the persistent, and I believe incorrect, theory that humans came to the Americas at a very late date following the humanization of the Old World.

---

[132] Ruth D. Simpson, 20 *Anthropological J. Canada* 2 (1982), 8.

[133] Virginia Steen-McIntyre *et al.*, 16 *Quarternary Res.* (1981), 1.

THE NEW HUMAN BEING

Upon a probable mutation, which has been described, the hominid was subjected to a general instinct-delay that left only "lower-level" and instinctive operations largely untouched (but not unreachable). The instinct-delay can be termed depersonalization, which was the first feeling of homo schizo, to be promptly succeeded by a splitting of the mind into multiple entities that ultimately became a typical human poly-ego. The depersonalization aroused the new creature to a high level of fear, a general anxiety, an existential fear, integral to its being, ineradicable. The response to the fear was a grasping for control of the selves to reestablish the former hominid consciousness and its instinctive nature. This was also permanent. The human was marked by a mania for control.

The control-mania could not stop with the selves, because the selves did not stay with the body. In the struggles among the *personae*, the whole world was embroiled; a splatter of displacements occurred. Streams of affect or identifications were ejected, with attachments ensuing, minimally at the level of attention. Attention extended to habit and to obsession and to a sensing of property, all being mental strategies to fix upon objects to control, thence relief of anxiety. The 'return on the investments' in real or sensed or illusory affect consummates the transactions, no matter whether with people, objects, or spirits; the transactions could be termed, also, projections and retrojections.

The outcomes of this unceasing and uniquely human transactional process are numerous. They can be grouped into:

a. Perception and attention with typical overlappings into perception disorders, hallucinations and illusions.

b. Thought, logic and analogy, moving into rationalizations, delusions and thought disorders.

c. Selective, recallable memory, often employing amnesia for fear-reduction.

d. Emotional ambivalence respecting all persons and things, a mild anhedonia and general negativism, anxiety-freighted, as distinguished from the hedonic animal.

e. Aversion or the non-acceptance of apparent *prima facie* resolutions of human relationships, including paranoia with its fearful denial of retrojected affect and the substitution of alternative hypotheses of threat.

f. Psychosomatism, the stressing of the body to achieve higher control levels, often with healing and destructive effects.

g.   Guilt and punishment, ranging among all persons, objects and spirits to discipline and erase fear.

h.   Discipline and work as an outcome of attention, habit, and obsession.

i.   Drug addiction for anxiety-therapy and orgiasm.

j.   Anxiety, which continually presents problems for solution, and, when overabundant and impractical, engenders neuroses, neurasthenia, epilepsy, and depression.

k.   Internal speech, the coding of information bits and sets, in time and space, for quick retrieval by association or for computation, including speech disorders when pieces of code are compulsively expelled as speech.

l.   Language, public speech, to signal and control the outside world, real and delusional, using internal code elements that others agree upon.

m.   Sublimation, the elaboration of symbolic activity in a low-anxiety area of displacements.

n.   Basal activities closely paralleling earlier primate behavior, such as eating, sexuality, mother-love, aggression, and fear-flight resulting from immediate threats, except that these activities are continuously subject to uniquely human interventions.

In the outpouring of his new nature, the proto-human thus exhibited new methods of handling large portions of the range of animal behavior. He could think about, talk about, and do something about a world of problems of which his ancestors were unaware. He would give a new aspect to all the ordinary activities of the earlier hominid. However, if eternal 'angst' be considered as a cost, the new person paid heavily for his virtuosity.

CHAPTER V

CULTURAL REVOLUTION

In dreaming, the human brain works fast, conjuring plots and actions that would be not only physically impossible but also temporally prolonged. Persons who have narrowly escaped an abrupt death sometimes exclaim, "My whole life passed before me in an instant." Many creative artists and inventors, whether in the physical or social field, refer to their visualization or conceptualization of a total product in a moment of "intuition."

Such occurrences point toward a theory of cultural hologenesis: if human, then holistic thought; if holistic thought, then holistic behavior; if holistic behavior, then collective instant culture, or at least a culture that develops as rapidly as the acting out of dream and thought sequences can be managed.

A culture - a group mode of mentation and behavior - arose promptly with homo schizo. Just as man became psychologically holistic upon his origination, so did he become culturally holistic. Human culture was global from its beginnings. Culture was schizoid and remains so.

The expansion of homo schizo geographically and culturally proceeded rapidly. Three hundred people, the number achieved in the first fifty years by the scenario of the last chapter, could, under optimal conditions, reach into the billions within a thousand years. Some millions probably did breed. His spatial movements, again if under minimal constraints, could carry him in ever-widening circles to the farthest points of the globe. Like population, spatial occupation probably did proceed exponentially.

For reasons given in my study of *Chaos and Creation,* it is unlikely that the point from which he was launched upon the conquest of Earth and its denizens is presently meaningful; the continents and the aquatic basins have shifted. His point of origin may be set at present-day zero degrees latitude, zero degrees longitude, without contravening any mass of evidence to the contrary. Neither the Iranian Plateau nor the rift valleys of Africa are any longer candidates for the spot. The mouth of the Mediterranean and the Caribbean Sea, if these were joined instead of being separated by the Atlantic Ocean, would be a likely homeland, but to argue the issue farther than we have done in the last chapter would lead far afield.

PROTO-CULTURE

The question is, how could homo schizo, granted his rapid increase in numbers and territory, accomplish the acculturation of which we speak? We know something of his psychology. How would this originate a culture?

What we have to demonstrate is that within a century or two, the major structures of culture would be necessarily, recognizably, and irreversibly present wherever the human race was found. These would be implicit in any one of many things that must derive from self-awareness: speech, tools, voluntary organization, religious symbolism, new constructions, movable property, fire tactics, time-factoring.

The first culture was a set of wild moves in all directions guided by displaced instincts and an intense need to stabilize the psychic world. It was like the output of a newly designed computer that had to be newly programmed to process data that had to be freshly gathered in order to satisfy the new program.

Usually the search for culture begins with a search for tools, because tools can be hard and enduring, and because they exhibit a deliberate human effort to command materials to effect a purpose. We should acknowledge first, though, the inevitable and greatly convenient built-in tool kit of a human. The first human was a tool-user whose body was his portable tool-kit. The hands of the ape are not put to many of their human uses. The human made tools of his fingers, hands, arms, feet, back and shoulder muscles, tongue, spittle, voice-box - indeed of all his senses and organs that he could command. Even today in a highly technical society where there is 'a tool for every purpose, ' the built-in tool-kit is continuously in use in ways far exceeding the imagination and capabilities of the primates. One can indeed conceive of a culture without artifacts. But in reality man must go on to make other tools. He has no choice.

Like man's anatomical tools, his mechanical tools are projections of nature and analogies to it. They exhibit a sense of the future and represent the obsessiveness of humans. The tool is pragmatically rational if, in addition, it is functional (efficient) and conserves resources. A tool, then, is a socially transferable physical object believed by its user to confer a larger control over the world than he could otherwise achieve. A mechanical tool is a type of social tool, also, and there is some merit to defining a social tool as an organization of other people believed to add to the user's control of the world.

Who ports a club, supports a culture. He remembers to carry it, and foresees a use for it; so he has memory and foresight. The club is a versatile tool against living things and obstructions; it extends the arm and gives leverage. It has to be produced; a skill is involved, so we have *homo faber*. It is one's own, so we have a property right. It is a coercive threat; it is a sign of fight more than flight, so it communicates a sign of power and authority. As the "batons" of upper-paleolithic man evidence, the club converts readily to a work of art, employing symbolism of lines, geometry, living things, and carved depictions of the phallus. The club reaches into the sky to connect with shooting stars and comets, as in the snake-entwined rods of Hermes and Moses.

Thus the simplest tool, the club, represents the major areas of human interest: skill, subsistence, economics power, safety, authority, sexuality, religion, and aesthetics. It is required, however, that it be carried. If we knew when the club was first carried, we would have a sound basis for fixing the gestalt of creation in time. Alas there is no earliest club; wood rots quicker than bone; we have only the aforesaid early bone batons. We have practically nothing belonging to the earliest man, nor ropes, skins, bamboo constructions, and so on; all subtle evidence is gone, leaving chipped stones and stone mounds.

Our statements, such as these about the club, must be highly speculative, anchored mainly by a theory of human origins and nature, and by retrojections of tribal practices today. Tool kits of different cultures might be counted. Leroi-Gourhan has estimated the oldest cultures, the Acheulian, to have possessed 26 stone tools, the Mousterian Neanderthals to have 63, and the succeeding modern type to have 93. These kits do not include all of the tools by any means - not skins, vines, ropes, gut, shells, bamboo, leaves, clubs, wood levers, wood slides, bones, hair, fur, paints, glues and so forth. E. H. Man's survey of the isolated and simple-living *Andaman Islander* a century ago revealed no more tools of the stone type but more made of the material that would have been destroyed by time and nature. Such tools might raise the given numbers by a factor of five, giving 130, 315 and 465 which, averaging (for who can say what determined the ratio in each case), gives some 300 tools in earliest known human cultures. Then add the tool chest in the human body. We can take it for granted that the earliest human who used *any* tools, used *many*.

With such material uncovered from, or imputed to, paleolithic man, a world of intellectual, emotional, ritualistic and mundane variety can be contemplated. An engraved ox rib from Pech de l'Azé was called Acheulian and dated at 300,000 years by F. Bordes, and in 1982 Pietro Gaietto published in Genoa a treatise on pre-historic sculpture.[134] There he moves the earliest artistic works of mankind "a million and a half years" back to the pebble culture of australopithecus and homo erectus. He argues that the earliest busts and menhirs are as decipherable as the earliest utensils, and exposes abundant evidence of artistic mentation in material of a type hitherto disregarded and cast aside by paleoanthropologist.

Appearing first in what may be artificial modifications of naturally suggestive stones, they develop successively in pre-Neanderthal, Neanderthal, and homo sapiens excavations. Working independently, L. G. Freeman and R. G. Klein, University of Chicago anthropologists, announced a year later the discovery at El Juyo (Spain) of a sanctuary containing a probable altar, weapons, house-hold tools, animal relics, and a stone sculpture. The sanctuary was dated at 14,000 B. P. and the bust depicted a two-faced creature, half smiling man and half cat. It resembles a number of Gaietto's sculptures.

Gaietto's controversial findings conform to my theory here, lending more evidence 1) of the humanness of the hominids, 2) of cultural hologenesis, 3) of a persisting interest, amounting to an obsession, in two-headed and two-faced persons, that may denote wonderment over the self-awareness of homo schizo, 4) of concurrent cultural and physio-psychological human genesis, and 5), although he does not question the conventional long-term chronology, of the cultural homogeneity of paleolithic beings and therefore of a short elapsed time since humans quantavoluted.

Even though he believes in darwinian gradualism in human development, Andre Leroi-Gourhan can say of his study on prehistoric religions that

> Man, from his formation up to our times, began and developed reflection, that is, the ability to translate the material reality around him by means of

---

[134] A. Marshack, *Amer Sci,* March, 1976 and *Curr Anthro* (1976) 278; Gaietto, *Prescultura e Scultura Preistorica,* E. R. G. A.: Genova.

symbols... There is no good reason to deny to paleolithic man a preoccupation with mystery, if only because their intelligence, of the same nature if not of the same degree of *homo sapiens*, implies the same reaction in the face of the abnormal, the unexplained. Here, facts exist, many of them, which show that from his first moments, *homo sapiens* (or his immediate predecessor) behaves like modern man. The indicators involve not only religion, but also techniques, habitations, art, self-adornment; they create, by contrast with that which precedes, an intellectual ambiance in which we recognize ourselves at first glance.[135]

He is saying that modern man has been basically unchanged from his beginnings. But the beginnings for him go back millions of years and we assert that the evidence of a long period is almost entirely wanting.

S. A. Semonov, in his work of 1973 on *Prehistoric Technology*, attempted an analysis of the stages of technological development followed by mankind. He perceived seven tendencies, which he believed to have followed one another over a long time. First a manufacturing process was invented to reduce the angle of edges on stone. Then smoother blades were evolved to reduce friction. Next, the mechanical power of tools was increased by elevating the amount of force that could be applied to the instrument. Steps were then taken to increase the rapidity with which the tool can be exercised while working it. Specialization was afterwards introduced to evade the limitations of a general tool and accomplish better the foregoing processes. Later, the physical-chemical properties of the instruments were enhanced by using fire, sunlight, and water to alter the properties of rope, wood, and bone. Thereupon, abrasives and saws were invented to increase friction, and the pestle and mortar were employed to pound materials.

We note that the principles of force, involving portage (pushing and pulling), the lever, elasticity of matter, gravitation, and chemical combustion, were incorporated in the processes. Too, wind, sun, and fire were used directly to play upon the materials and convert them. Animals, furthermore, were induced to dig, carry, and turn devices, much later on, it is thought, and animals too were exploited, as with bird-eggs and bees honey.

Yet there is no rigid requirement that these inventions should follow one another in all cases, or, if they did, that they should not have followed one another quickly. It is the counting of time that lends an evolutionary atmosphere to the proceedings. A more rapid counting, on quantavolutionary theory, would accomplish the same developments in several hundred years. Much depends upon intensity of motive and self-awareness, once the time element is laid aside. The concept of the gestalt of creation, we have argued, supplies such strong motivation and awareness.

---

[135] *Les Religions de la Préhistoire*, Paris: Presses Universitaires, 1967, 6, 7, 146-7.

We go further and claim that in his first years on Earth, homo schizo must have achieved much in the way of tools and culture. It is safe to say that, if at all human, that is, if self-aware, hence finding many objects and animals of interest and striking for control of the world, homo schizo would in short order arrive at a complete culture-kit.

I have already shown that, to paraphrase Bonaparte's remark about bayonets, a self-conscious person can do anything with a club plus sit on it.

Also, you can digest any organic material that you can find and eat, raw or processed. Processing includes to stew, heat, bake in ashes or sun, salt, soak, pound, powderize, and pre-masticate.

You cannot gather plants without noticing that they grow from seeds, and that seeds and bulbs are edible, and that time after time your favorite location will renew itself.

You cannot chase animals without catching their young, then raising them until they are ready and needed for food. (Modern women have been noticed to nurse suckling pigs, until these can eat other food.)

You cannot have a garbage pile without observing that rodents, birds, and tasty insects feed and breed there.

You cannot handle fire without preserving it, using it for roasting, and being 'spiritualized' by it.

In skinning an animal before eating it, which can be done with one's hands, though a sharp rock is better, you cannot help but sit on the skin or use it as a muff or blanket or haft.

You can frighten and inspire responses by hooting and whistling, and whipping branches in the air, and if you frighten other beings and they you, can readily try to impress inspirited locales, like caves and sky, where you imagine there must be live things, to keep them from frightening you.

You cannot gather eggs without finding young birds whose wings you can break and which can be kept in a loosely covered hole until grown.

You can get agreement with others by recalling and using sounds in common, and can convey known sounds from one person to another from one day to the next or one place to another - a message.

You can model your indecisive behavior on your remaining instinctive behavior and animal behavior, unknowingly setting up the paradigm of logical and pragmatic thought about causes and effects.

Should not these necessary immediate implications of proto-human brainwork be incorporated into appraisals of earliest man? No evidence contradicts the statements; why, then, should a creature be put to climbing the rungs of a ladder for four million years or forty thousand years, for that matter?

Probably the ideology of classical anthropology was at fault. In order to "discover" proto-man, the amateurs ventured forth among the most "savage" tribes. The most "savage" would be the poorest in property (the heyday of the bourgeoisie was then) and the "simplest" (rococo-type art permeated the Victorian age). So the students referred to the peoples who were "hunters and gatherers."

Instead of penetrating into and evaluating the mentation of these peoples, explorers and reporters placed them into the category of primeval man, who had to be one step above the apes and who had just climbed down from the trees. Probably there was in this theory, such as it was, an element of ethnocentrism. The British geologist Ager has noted that the nomenclature and systems of rocks in the world have had a suspiciously prominent presence in the centers of the old British imperial posts and routes. The British invented and dominated much of early anthropology, too. A joke as hoary as Queen Victoria went, "One Englishman a hunter, two a dress-up dinner, three a club." The "most-savage" nomadic hunter-gatherer (the women gathered) was a wish-fulfillment; Tarzan, son of an English nobleman, was back among the apes.

To the contrary, proto-human had very soon a culture that was as schizoid as he was and held the essentials of most subsequent discoveries and institutions. He invented as he moved through the world, and the news about, and practice of, culture moved with him. Settled and mobile communities existed, tied into the ecumenical culture, kept posted by eccentric wanderers and by group encounters.

LOST MILLIONS OF YEARS

By extensive comparisons of primates and mammals, Robert Martin has positively related basic and active metabolic rates to body size, then again body surface with brain size.[136] Brain size and body size are also positively related. Man's enormous brain is partly accomplished in embryo and partly post-natally. The big spurt after birth, when coupled with the very small human litter, typically one infant at a time, leads Martin to believe that this relative human abnormality depended for survival in the process of natural selection upon the persistence of a stable natural environment and ecology. In our terms, this would imply a denial of the need for a high reproduction rate as insurance against catastrophe.

---

[136] Roger Lewin, "How did Humans Evolve Big Brains?" 216 *Science* (May 1982) 840-1.

The human reproduction rate, however, is compatible with catastrophic conditions; it is still exponentially high. Furthermore, only because it is working humanly and not because it is large, it is a pragmatic or "rational" insurance against catastrophic obstacles to survival. Therefore we would discount the meanings that have been offered of his correlations; there must be some significance to them, but not the one for which we are searching.

The catastrophized human mind is itself proof against catastrophe. The human, it appears to us, must have grown a larger body and brain, and heightened its metabolism, and lengthened its training period because it was already human. Stated simply and crudely, the human wanted to overcome its disadvantages and extend its controls, and did so - genetically by breeding, psychologically by practices and ideals; it invented the gods and imitated them.

Population growth rates present no obstacle to a quick diffusion of mankind. They are an exponential phenomenon. An amusing calculation recently gave to Charlemagne's fifteen children of the ninth century some 255 billion contemporary descendants, a hundred times the world's population today. (Obviously heavy intermarriage occurred continuously since his day, especially inside France.) Then, the genealogist said, Attila the Hun several centuries earlier made his presence felt in what became the kingdom of the Franks, and Charlemagne had to be descended in some part from Attila, by statistical calculation. Which would permit finally every modern Frenchman to claim descent from Attila. For that matter, many of us may descend from a fecund cousin of "Lucy," the australopithecine who perished in the ash wastes of Afar.

Anyone in the world can play a similar game. Populations, human groups included, repeatedly expanded and contracted like an accordion, in the passage of centuries. Today we are impressed by expansion. The people of India number over 700 million, twice the population of 35 years ago, and pressing hard upon the means of subsistence. Yet they are projected to double in the next 32 years to 1,400 million. A quantavolution, whether deliberate or disastrous, is foreseeable.

Man should have reached a comfortable Neolithic level of culture within a thousand years of humanization, and stayed there unless general catastrophe intervened. The Neolithic is universally acknowledge to have been an across-the-board human culture with all basic practices, institutions and techniques invented and in use; it was certainly in being everywhere 8000 years ago. Could man have been fully potentiated and activated by mutation - i.e. physiologically complete as a human - but not have behaved so as to develop his mind and culture except very slowly and incrementally?

If so, then what was retarding him, keeping him for periods of first millions, then thousands of years, from making progress towards the new Stone Age?

Might it have been perpetual dietary deficiencies? But the diet of the hunter-gatherer is excellent.

Might it have been perpetual warfare? But war has incited invention and cultural diffusion throughout history. Moreover, war may not have been continuous.

"Neo-malthusianism" and birth-control among the race as a whole or among the intelligent would be implausible.

Might it have been the difficulty of first inventions, as opposed to secondary ones? The lever, the spring, the knife, the bucket, the garment, the overhang, animal training, the advent of springtime seeding? Are these inventions which would be taking trillions of man-hours?

Continuous plagues of types known and unknown today? A generally stupefying plague is unknown.

Might it have been a world catastrophe (climatic, fall-out, solar black-out?) These would endure only briefly.

Were there recurrent global amnesias from a stupefying and dizzying electrical condition of the Earth? This is conceivable.

Might it have been frequent devastating natural disasters? Like war, disaster teaches.

Was there a catatonic fear of change - a frozen taboo against change? Changes are eventually forced, and taboos do not block all avenues of development.

Might life have been simply too easy, hence *dolce far niente?* Life (see all above) is never that sweet; and recall his eternal *angst.*

Were men too few or isolated? Not knowing about each other? Contra-indicated.

Perhaps they could not organize a division of labor? But the potentially *useless* would have a desperate motive to make themselves useful, to avoid being discarded.

"Whatever the reason, the primitivity of many tribes today shows that men do not progress except for reasons which we do not understand." But they succumb to new temptations right away - horse, ax, gun, tobacco, sugar. Further, as we argue, primitivity may not only be a mistaken idea; it may in any case be an actual short-time, youthful phenomenon. If "primitives" act young (" the childish peoples" some early anthropologists called them condescendingly) it may be because they are young, and so are we all.

I cannot completely dispose of all of these objections here. Merely to phrase them, however, disposes of many. The very nature of homo schizo as a restless, anxious, control-seeking creature answers them. The most troublesome problem, it seems, is a possible variant of the events that produced homo schizo: if a subsequent new constant of a gaseous or electrical character were to be introduced into the atmosphere, mankind might be numbed or frustrated mentally for a hundred or a million years, a prolonged Tower of Babel effect, one might call it. By a worldwide alteration of electrical fields, the human mind would be incapacitated for consistent and routine solutions of problems; it would be amnesiac; it would be fibrillating excessively and continuously. Or, conversely, the mind would be deprived of the hormones and gases required for all except quasi-catatonic operations; mankind would be a sleep-walker for millions of years.

Evidence has been already presented to show that these lost ages have not occurred; they never existed. Hence, this possibility must be preserved only to defend the theory of homo schizo in the event that long-term time reckoning turns out to be correct. I shall continue, therefore, my analysis, tending to show that human culture has not been slow in developing, but, to the contrary, rapid.

Scholars have sometimes wondered at the long ages of mental stagnation. Thus, J. Hawkes remarks, "That a tradition could continue with only slight changes of essential style over a period of between twenty and thirty thousand years, which is what our present chronology suggests, seems today almost incredible." [137] If this scene of the Upper Paleolithic is incredible, what then of the hundreds of thousands, even millions of years, of changelessness going before?

Thus Sol Tax comments upon "the universality of the material characterizing the East, on one hand, and Africa, Europe and India, on the other," and how their artifacts span "four-fifths of the quaternary period" with practically no change, and "a socio-cultural reconstruction of the Sinanthropus cultural material would be mathematically the same as that made for the Australopithecines." He concludes that "Certainly, the stability of attainment and the lack of change cannot ever be taken as characteristic behavior of Homo sapiens as we know him, and we must look closer to home for our first representative of Man." [138]

In effect, he is saying: deny man exists, as long as he is not developing for long stretches of time. Instead, Tax should be challenging the time-

---

[137] *I Prehistory,* Part I, 280, New York: Mentor, 1965.

[138] "Primitive Man vs. *Homo Sapiens,"* in A. Montagu, *The Concept of the Primitive,* N. Y.: Free Press, 1968.

clocks. His position seems all the more uncomfortable inasmuch as he has acted as a leader in bringing the public to realize that primitivity is a pejorative term and unjust to the mentality and culture of 'primitive' peoples.

TRIBES, CIVILIZATIONS, AND TIME

I prefer the term 'tribal' to the world 'primitive': it is less misleading. Tribal cultures are not young; they are as old as the oldest modern culture. All cultures are equally old, so far as one can tell. The tribal culture holds a stronger illusion of special gods and heroes; it claims common ancestry; it speaks a special language; most of its transactions are inside the tribe; and it has not been accommodated to a greater society. These conditions are disappearing; few tribal cultures are left outside the great society.

Until recently, many tribes were 'resting' in the Stone Age. This is a mechanical and psychological judgement, not an ethical one. As Jules Henry and others have explained, their "psychic unity" is complete.[139] When a culture achieves some tolerable mastery of its individual and collective minds, there is little incentive to change unless it is ravaged by nature or conquest. A scarcity or profusion of artifacts is no proper criterion of the humanness or human development of a culture. Until this century, a village with its farmers on some Greek islands would possess few artifacts, and its church would be scarcely more than a shaman's hut in Central Africa. If the Greek and African villages were compared with the Shandridar village of ancient Iraq or an early community in the Basin of Mexico or classical Tiahuanacu one could not argue conclusively that the later were more evolved than the earlier, and one would find perhaps similar difficulty in appraising their outlook and mentation.

The great civilizations began to appear about seven thousand years ago with commerce and conquest. There have been perhaps fifty of them. They take about three hundred years to gestate and last for a millenium before handing themselves over to another civilization as with the Incas, and/ or dissolving, as happened with the Roman Empire. During this time, and counting the component cultures from which they were amalgamated, there may have existed about 20,000 cultures in all.

It is difficult to 'put a tribal culture back together again' once it has been absorbed into civilization. Sometimes a tribal culture will remember having been ruled distantly but not tightly or absorbingly. It shows almost no signs of having been included in a bygone civilization. Therefore it must have

---

[139] "The Term 'Primitive' in Kierkegaard and Heidegger," in Montagu, *op. cit.*, 89.

existed by itself since the beginning of human time, or since it split off from a tribal aggregate at some time in the past to form a related unit. The fission would have occurred because of natural catastrophe, flight from a growing civilization, internal disputes, or overpopulation and emigration.

It is unfortunate that all of these statements must be conjectural. Yet their thrust is unmistakable. There has not been enough time since the beginning of human culture for all tribes to have experienced participation in a major civilization - except for the ecumenical proto-culture to which all peoples must originally have belonged. In this case, the demography of cultures would imply recent human origins and support the theory of cultural hologenesis of homo schizo.

Elsewhere I have defined a 'memorial generation' as a unit of fifty years that would span the age of the oldest story-teller and the youngest attentive listener of a group. It is about three times the length of a reproductive generation. Some current estimates, using long-time reckoning, have human culture appearing, bit by bit, of course, for from 50,000 to 5 million years. Here we estimate that one thousand years (20 MG) is enough; and 260 memorial generations (MG), 13,000 years, is enough for the history of mankind. Fifty thousand years give 1000 MG's, and 5 million years allow for 100,000 MG's. Unless the human mind developed finely, bit by bit, with one tiny innovation following another, the human could not consume so much time so unprofitably. And what was directing this incrementally minuscule evolution? And if it burst into quantavolution in the Upper Paleolithic, what caused that event to occur?

If it were not for the accepted methods of reckoning time, scientists would probably have to agree that a hologenesis of both man and culture is logical and recent. To hallucinate further, if Solon of Athens had called on a panel of experts from Babylonia, Iran, China, India and Mexico, as well as from Greece and Egypt, in the sixth century B. C., all would have told him that man's history was short, at least since the last great catastrophe. But belief is firm in the tests that report long times for the early fossils and relics of man and life generally, and claim a long, slow ascent.

A century ago, when time reckoning was governed by our type of speculation, by the fossil record, and by the apparent ages of sedimentary rock strata, time measures were easier to assail. Today, radiochronometry lengthens human time and fixes it by elaborate chemical tests, the most critical of which are the radiocarbon dating and potassium-argon (K-A) dating to which I have made reference above.

Both tests are striving for validation in the crucial middle times between 10,000 and 100,000 years ago. It is expected and hoped that they will close this gap. Meanwhile the K-A test can be used to support very old ages for what appear to be human remains with artifacts; and the 14C test is keeping the Upper Paleolithic age far enough back to support impressions of a very gradual human cultural development. I have given elsewhere my reasons for disputing the validity of 14C beyond 2700 years, for regarding the K-A test as quite unreliable, and for questioning most other chronometry. (Chaos *and Creation*, chap. III)

Here, I do not treat fully these tests, because the theory of human and cultural hologenesis is independent of the time-tests frame. Hologenesis could have occurred 13,000; 50,000; 200,000; 1.5 million or 5 million years ago, except that in all of these cases, an incredible amount of human history is missing. Perhaps we should hope to find it, cheered on by the late reports from micro-paleontology that have added a billion years to the two billion year age of life on Earth (but brought the age of life and the age of the Earth itself uncomfortably close to one another).

For those dates that are beyond 50,000 years, one might postulate a limited jump in human and cultural evolution, leaving a final large jump for the Holocene boundary. That is, some hominid, perhaps not even a human ancestor, could have chipped a stone, with nothing else on his mind. Given my analysis of the club-wielder, I would not know how to explain this activity. It would not be modern man, but a different species. I find this solution easier to tolerate than a gap of millions of years between a true man, a chipped stone, and the Upper Paleolithic-Holocene periods.

MAJOR DEVELOPMENTS EVERYWHERE CONTEMPORARY

Not only did primeval man quickly achieve a world-wide protoculture, but the next age, so far as we can tell about ages, reveals increasingly a panorama of cultures of equal status around the world. To distinguish this age from *proto-culture*, let us refer to it as *neo-culture* and think of it as merging the Upper Paleolithic, Mesolithic, and Neolithic developments.

Legend usually does so, and in a way so do the most ancient texts. Man is created, he is savage but human, he is given gifts of all arts and crafts, he lives in a golden age. He is destroyed, and a new age follows. If a new panel of experts were called, this time by name, say High Priest Aaron, Akhnaton of Egypt, Solon, Hesiod, Plato of Greece, and Ovid of Rome, they would probably agree to this and add much illuminating detail.

If the French scholars of the years of nationalistic jealousy were not intent upon showing the great age of advanced culture in France, they might have assigned the cave art of the Dordogne to the time of pre-dynastic Egypt, 6500 years ago. The hot breath of tourists damaged the Lascaux paintings in a few years; before then, neither the ancient users nor the dozen thousand years of quiet cold damp were sufficient for their destruction. Nor the great climate changes that drove off the cave people and the large animals. Of their style in general, Leroi-Gourhan writes, "The nature of the paintings does not seem to have varied from -30,000 to -9,000 years before our epoch and stays the same in Asturias as on the Don River."[140]

Some 21,000 years of the same genre of painting. And by now he would have to say -- from England to Siberia. The older genealogy hardly justifies the assigning of ancient ages; it all was suspended by the thread of thriving ethnocentrism, until geophysics advanced a radio-carbon test for charcoal and bone.

From A. F. Spiess' *Reindeer and Caribou Hunters* (1979), we are permitted the notion that protohistorical North American hunters and Paleolithic hunters of Southwestern France (Abri Pataud) had similar relationships with their prey, despite the numerous different cultures in each setting and within the settings, over a passage of up to 35,000 years. The social adaptation of humans to animals suggest common behaviors persisting universally (relative to the ecology) over long time spans.

The mode of life of the 'hunters' of the Upper Paleolithic, which has now been extended beyond France as far as Siberia and through the Sahara possibly down to Southwest Africa, and very lately to England, may not have been the exclusive life of the times. The caves themselves were not for living. As far as one can tell, they were modelled on the cloudy vaults of heaven and the mysterious depths of the womb. The passages and chambers were artistically organized for stages of religio-clan initiation. The bison was the central totem animal.

Living, for the hunters, was outdoors, or in temporary huts, or under abris which could shelter them against the elements. They must have been connected with settlements, where the women and children and animals would have stayed. One cannot examine their artwork without grasping that at the very least they would be living in the style of the North American Indians before 1600 A. D. Repeated devastations and heavy sedimentation and sinkings have obliterated practically all traces of their villages and gardens, and perhaps major civilizations as well.

Generally the domestication of animals has been placed in the period 7-9000 years ago.[141] A claim is now advanced for domestication near Nairobi in East Africa at 15,000 years ago. In Patagonia, whose natives are looked upon as exceedingly 'primitive', men long ago captured, confined, fed, killed,

---

[140] *Op. cit.*, 85.

[141] R. Protoch and R. Berger, 179 *Science* (19 Jan. 1973).

and ate the giant ground sloth, now extinct, the extinction perhaps occurring when 70% of the great pleistocene mammal species disappeared.[142] This will certainly confuse the picture.

Meanwhile agriculture seems to be moving backwards in time reckoning. C. Niederberger finds Mexican sedentary economies with a mixed agricultural-gathering-hunting base around 8,000 years ago. "Artifactual and non-artifactual evidence from the lacustrine shores of the Chalco Basin already suggest the existence of fully sedentary human communities in this region from at least the sixth millennium B. C."[143]

San Pablo (Ecuador) corn kernels embedded along with associated corn designs on pottery in deep cultural remains "show a heavy agricultural population between 200 to 4000 B. C." (using 14C tests with bristlecone pine correction). These high flood plain sites are called generally the "Valdivia" culture. They are definitely not of Japanese culture type, as may be some other early discoveries of the same region. Agriculture was known throughout the world in Neolithic, and perhaps much earlier times. One may ask whether agriculture, which is not an easily diffusable set of inventions, was not practiced *in embryo* during the first ecumenical culture of homo schizo. Southeast Asia and Asia Minor are emerging with concurrent early dates.

We can quote Henry T. Lewis:

> A search for the various stimuli to domestication should not involve looking for those factors which led man to discover agriculture; rather it should involve learning about those factors that made agriculture a necessary alternative in human adaptations, first as a complement to hunting and gathering, and later as a substitute for it.[144]

In pre-European California, hunting and gathering competed successfully with agriculture,[145] for example. And, again, Lewis writes: "Domestication would have begun not as a 'revolution' but, rather, as an attempt to extend and stabilize the existing subsistence strategy."

---

[142] C. M. Nelson, *N. Y. Times*, Aug. 27, 1980; A Smith Woodward, 15 *Natural Sci* (1899) 351-4.

[143] 203 *Science* (12 Jan. 1979), 140; R. S. MacNeish, "The Origins of New World Civilization," 211 *Sci. Amer.* (Nov. 1964), 29-37.

[144] "The Role of Fire,.." 7 *Man* 2, 1972, 217.

[145] Sahlins et al., *op. cit.*, 77-88; *cf.* MacNeish, *supra fn* 7, p. 36, where, despite rich variety of domesticated foods, only 10% of food supply came from them (ca. 5000 B. C.).

Here he is saying what I earlier implied, widespread natural disasters may have driven humans into agriculture, away from the more convenient and satisfying life of the hunter-gatherer, 'just as the Bible says. '

As for today, the same group of anthropologists agree that "it is merely a matter of time before all the cultural systems of the world will be different variations, depending upon divergent historical experiences, of a single culture type."[146] This exemplifies their law of cultural dominance. But it also casts doubt on any great antiquity for culture, hence for man.

Baker comments on the situation concerning prehistoric botanical domestication and diffusion, saying "Why is it that the ancestors of domesticated plants are now so rare (or even extinct)? It is hard to see how domestication of *cucurbita* (squashes) would make life any more difficult for the wild species."[147] Might it be that man under catastrophic circumstances takes care of his plant seeds while the wild seeds are destroyed? Might man also preempt the best areas for growing the plant, thus handicapping the wild sort? And might not the wild plants have come from an isolated botanical niche whence they were transported around the world by men? All three arguments, especially the last, appear to be valid. They would point to an early, rather than late, date for agriculture.

It is not impossible, then, that much of the Upper Paleolithic and the Neolithic were merged, and throughout the world, too. As for the Mesolithic, this usually maligned cultural epoch is now receiving accolades for its own achievements. Nevertheless, it still hardly has boundaries to distinguish it as a period, and rather is sandwiched in between the two other periods to fill the greedy stomach of time.

To support the foregoing hypothesis, it is well to stress again that many tools bridge gaps of thousands and even 'millions' of years between different epochs, leaving one to wonder at the marvelous resiliency of the stone age peoples who otherwise appear to reject invention. So that, for instance, Neanderthals executed works of art.[148] They also garlanded their dead with flowers in Northern Iraq,[149] implying an interest in horticulture, as well as religion. Neanderthal Mousterian styles of stone-working are found in Magdalenian deposits.

And a single Greek cave assigns one chipped stone from Upper Paleolithic to early Neolithic, an adjoining stone from Early Neolithic to

---

[146] Sahlins et al., *op. cit.*, 92.

[147] In Riley, Ed., *Man Across the Sea, op. cit.*, 441.

[148] Walter Matthes, IPEX; *Jahrbuch für prähistorische and ethnographische Kunst*, 1963.

[149] Ralph S. Solecki, "Shanidar IV, A Neanderthal Flower Burial in Northern Iraq," 190 *Science* (28 Nov. 1975), 880-1.

Late Neolithic and another from Middle Neolithic to final Neolithic; Magdalenians use Mousterian tools, etc.[150] And, once again, MacNeish, working at Teotihuacan sites, assigns one type of uniface flaked stone implement at 10,000 years of age and finds that it continues to 6300 years ago. Just before this last time, another type of implement picks up and carries on until 800 years ago. Nine thousand years are spanned by two implements.[151]

ECUMENICAL CULTURE

There was in the beginning one human race, one language, one culture. The contrasting hypotheses seem to be losing vigor. That many cultures around the world originated independently implies that men scattered around the world and only then started up cultures from a delayed time-fuse in their brains.

Despite the tenacity with which this idea grips many people, it would appear absurd, unless one believed at the same time that humanization occurred immediately in consequence of an atmospheric change that affected the brain with some uniformity everywhere, an idea that I have not seen expressed except in these pages, and here is included as a partial and fluctuating cause of humanization.

Since we cannot agree precisely when humans originated, for certainly most will accept Charles Darwin's series of insensible gradations in preference to my theory of holocene hologenesis, how can we fix a point for the beginning of culture? As I construe the conventional argument, it must assert that the ever-extending ladder of evolution contains many rungs, some of which are physical gradations and other cultural. When a physical rung - say a straightening of the spine - occurs, the lucky straight-backed clan is different from all other men until its trait overcomes their curved spines; but, meanwhile, some curved-back clan invents a bull-roarer, which gives an impressive sound, and this artifact begins diffusing among the curved-backs and the straight-backs, helping both to survive in competition with men of either type, as with animals. Hence, at any given moment in this long period of human evolution up to the present, one would encounter a dizzying number of intersecting circles of diffusing physical and cultural traits. Too, it is a competition of all against all. The number would be perpetually large, and uniquely combined at any point in demographic space.

---

[150] T. W. Jacobsen, "17,000 years of Greek Prehistory," 234 Sci. Amer. (1976) 19, 80.

[151] *Op. cit.*, 64.

So the mills of evolution by natural selection and mutation would have to be working very finely, very rapidly, and continually.

Inasmuch as this theory, perhaps exaggeratedly put here, dominates scholarly thought, all coincidences of cultural traits following humanization must occur by means of independent invention, or by adoption (that is, by diffusion from one or the other source or a common third source). Hence argument always centers around these two ideas and they have been flailing at each other in their boxing ring since the beginning of the uniformitarian orthodoxy a century and more ago. The additional contestants that I would sponsor here, out of a sense of sportsmanship, namely common origination in cultural hologenesis and common experience of general catastrophe, are barred from the ring.

It is easy to see why this prejudice should occur. With a very long evolution time, it is presumptuous, if not absurd, to believe that any culture trait possessing particular recognizable form could be part of a primordial culture. That is, such a namable and tangible trait cannot be very old. The idea behind the trait may be very old and represented in some now extinct forms and cultures as well as in present-day cultures.

For instance, the taboo against incest extending to first cousins is found here and there. These cannot be primordial but must be independent inventions, according to long-term evolution; they would be offshoots of a very ancient taboo against incest that may have conquered all cultures in some form by diffusion or independent invention at some time in the murky history of man. Freud's speculation that this taboo may have occurred everywhere by diffusion as part of a guilt reaction, also diffused, originating from the murder of the leader of a single primal horde, seems too close in time and has not been accepted by the orthodox anthropologist.

The prejudice against the arising of cultural traits out of similar experiences with a common catastrophe is also easy to explain. Such catastrophes have until recently been certified by astronomers and geologists not to have happened, or to have happened so long ago that they cannot have affected whatever it is that interests anthropologists or archeologists or prehistorians; hence no further consideration is required.

The literature of prehistory is otherwise rich in the assumed effects of climate, topography, and habitat upon cultures, deriving similar cultural and even physical traits from the similar experiences of men. Thus comets terrify all cultures. But this is explained as normal fear of unusual sights in the sky. The deluge is attested to by practically all cultures. But this is explained as exaggerated accounts of flooding and high tides. On the other hand the people of the north are blond because they need to absorb sun while the people of the topics are dark because they need to reject the overabundant sun. (It seems not to matter that the Eskimos and Lapps are dark, or that the great tropical forest scarcely illuminate the dark people in them.)

With all of this, there has until now been little chance of emerging from the source materials with even the beginnings of a division of culture traits as we conceive of them: elements that are assignable to times of common catastrophic experiences; independent inventions that came about owing to cultural peculiarities of given peoples with some parallels to be drawn from the independent inventions of other peoples; and innovations originating among one people and diffusing to others, whether in the wanderings after natural disasters and war, or in variously motivated migrations.

For instance, fire, which had been known to, and used by, hominids and other animals, would have been reinvented by mankind. "Fire was born when heaven and earth separated," says a Mongolian marital prayer. Fire - in its modern sense of something to be used multifariously, made and remade - was invented because the created human was terrorized by new intensities of fire, because the projected gods used fire in the skies and on earth, and because the new mind could remember its use and foresee its future utility. Credit for the invention was ascribed to a god and sometimes also to a god-hero who, partly man and partly god, could arrogate credit without displeasing the gods.

The earliest town plans were built according to a celestial model, and the planners were astronomer-architects. The conditions for planning were, again, an aware and awed human group, a sky religion, a skill in retrojecting and rationalizing a celestial scene, and then a science of measurement and construction. The orientation of the towns (Greek: *polis*) and temples followed first the North-South line of the Boreal Hole, a northern-most sky opening which happened in cloud-canopy times to represent the north geographical pole (from *polis*.) In less cloudy and in bright times, measurements that were derived from the old monuments and improved by stargazing, permitted the practice to continue. The Egyptian and Mexican city and pyramid orientations were North-South. In protohistory, East-West orientations became prominent because the sky-path of Venus was East-West, and finally the Sun's regularities provided the lines of true orientation for planners.

Macgowan's study of fifty early Mesoamerican towns shows modes at 70° East of North and 17° East of North, but several pitch from 1° to 21° West of North. The earliest are truest to the North.[152] La Venta was dated by Hutch (1971) at around 1000 B. C. and is oriented 8° West of North.

---

[152] In A. F. Aveni, ed., *Archaeoastronomy in Pre-Columbian America*, Austin: U. of Texas Press, 1975.

The changing orientations suggest that tilts in the axis of the Earth occurred from time to time; ancient man was never whimsical about orienting his towns.

A recognizably scientific astronomy is being sought farther and farther back in time. B. A. Frolov argues that an intellectual curiosity possessed early humans everywhere. The Russian counterparts of the Stonehenge monuments are at Lake Onega, and both are sky-directed religio-astronomical instruments.[153] The Pleiades are called the "Seven Sisters" by aborigines of Australia, North America, Siberia, and other ancient cultures. Petroglyphs that appear to refer to astronomical constants and phenomena are found all over the world; it may be mainly the prejudice on behalf of the 'evolutionary ladder' that forbids the assignment of many such carvings to the earliest age of humanity; in some of such cases the glyphs are found among the earliest ruins of a people or are the only remains discernible. That is, the hologenesis of mentation and culture derives support from the increasingly early assignment of scientific works.

Two thousand years after humanization, a large number of humans possessed self-awareness, religion and rites, planned towns, armed forces, a full range of stone and soft material tools, special occupations, domesticated animals and plants, and complex language. They entertained a range of aspirations that followed their time sense into visions of improved life; they created the rudiments of the highest ideals of later times: freedom from fear through knowledge, individual autonomy, conquest of the environment, storage against future hunger, and social cooperation. But the high energy forces of the gods permeated history, life, and expectations. Destructions were frequent, and catastrophes, such as they already dreaded, were to recur.

During the Saturnian 'Golden Age,' which was a single Neo Age, composed of the Upper Paleolithic, Mesolithic, and Neolithic, a wide circulation of traits occurred. Still a great many isolated groups, whose ancestors had survived the earlier catastrophes, continued to live apart. They became in many cases the so-called 'primitive tribes' of historical times, philosophically and technically undeveloped relative to newly organizing large central cultures. Later catastrophes added to the number of isolated units of culture.

Humans who are tribal in organization possess an essentially primordial culture. Among them are found well-developed languages in bewildering variety; they share not only linguistic principles but verbal roots with the great languages of the world; their attitudes toward language and symbols

---

[153] "On Astronomy in The Stone Age," 22 *Current Anthrop* (1981) 585; *cf* A. C. Haddon, 10 *Natural Sci* (1897) 33-6.

are proto-historical. Totemism is common; so also complex systems of taboo. Their religious astralism varies in extent and complexity. In comparison with scientized cultures, the succession of gods is less well described in legend, though the sky god (Uranus) is found everywhere.

Customs such as head and body deformation, and the couvade, that has the father imitating the pains of child-bearing, are similar in widely separated areas, suggesting an original universal community. Elaborated stone tools, advanced symbolic designs, ceramics, an attention to the North-South axis in monuments, the practices of circumcision, cannibalism, human sacrifice, flood legends, medical remedies, and a great many other practices and beliefs point back to humanization in the creative period, followed by devastation and isolation thereafter.

The primeval kit of humankind, the set of ideas and devices that the proto-humans gained by the gestalt of creation, seems less sophisticated than it really was. The voluntariness and self-consciousness infusing the cultural complex set it apart from mammalian products and organization.

Deliberate convocations and collecting of individuals into assemblies for planning, ordering, worship, and celebrating, accompanied by speech, symbolic gestures, markings, and rituals, also constituted part of the original cultural consensus - these in communications and organization. Planting, hunting, gathering, tool devising, storing - all operated from the collectivity extended through memorial generations - such were the practical activities.

Joseph Campbell puts our position here well:

> It has actually been from one great, variously inflected and developed literate world-heritage that all of the philosophies, theologies, mysticisms, and sciences now in conflict in our lives derive. These are in origin one: one also in their heritage of symbols; different, however, in their histories, interpretations and applications, emphases and local aims.[154]

AMERICAN CULTURAL ORIGINS

Alexander von Wuthenau, in his book on *Unexpected Faces in Ancient America, 1500 B. C.-A. D. 1500* [155] scans the literature on Asiatic, African, Egyptian, Semitic, and European presences in cultures and races of Central America and presents his remarkable album of stone and ceramic countenances of the stated peoples. Despite conventional theory, there

---

[154] *The Mythic Image*, Princeton U. Press, 1975.

[155] N. Y.: Crown, 1975.

seems but little question that the Central Americans were a mixture of human types long before Columbus arrived.

But further, the American race had its own primeval forms. In *Chaos and Creation*, as in the present book, I argue that homo sapiens schizotypus was present in the Americas from his very first period, and despite repeated general catastrophes held on there in niches of survival, and was repeatedly reinforced across the Pacific and Atlantic oceans, with artifacts and cultural practices to remind us of these occasions.

Although this thesis is not central to the present book -- because the theory of homo schizo can be argued on whichever grounds conventional theory chooses -- it has important consequences for early American studies. As I foresee the emergent issue, it is not rampant diffusionism versus carriage across the Bering Straits, but rather how much of the similarity among races and cultures came from the ecumenical period of homo schizo and how much was transmitted via long distances thereafter.

The case for diffusionism is building up. Some of the material advanced before World War II regarding Asia-to-America diffusion is summarized in Lord Raglan's *How Came Civilization?* (Chap. XVII). He placed the world ecumenical culture of the first civilization in the region of the Persian Gulf. More recently Betty Ebers has marshaled the evidence for Japanese to Olmec (Mesoamrica) diffusion, by sea.[156]

In another case, an authority on early Mesoamerica, Michael Coe (31) reports the "coincidence" from Needham's studies (1959, 407) that "the Maya astronomers and those of the Han Chinese worked with an eclipse calendar of 11,960 days." [157] The coincidence cannot be an accident, especially when one considers that the Mayans seem to have used 'solar mansions, ' like the Chinese, rather than a zodiac, to mark the progression of constellations, and, further, indicated constellations, in the manner of Han China, by circles connected with straight lines, which was not seen in Europe until 1785.

Acceptance and progress of pre-Columbianism are blocked mainly by uncertainties over the timing of intercontinental transactions. For example, Posnanski and Bellamy go beyond 15,000 years in reconciling Tiahuanacan (Bolivian) remain with Pacific Island and Mediterranean-Caribbean traits.[158] The Atlanteans range from 11,000 to 3,500 years ago. The Asianists for some time held to 12,000 by land and nothing by sea; then neo-Asianists

---

[156] "Yes, by Land, and No, by Sea," *Amer. Anthrop.*

[157] Michael Coe, in Aveni, *op. cit.*, 31.

[158] Arthur Posnansky, *Tiahuanaco, The Cradle of American man*, N. Y.: Augustin, 1958; H. S. Bellamy, *Built Before the Flood*, London: Faber, and Faber, 1943.

ascribed East Indian and Japanese contacts to materials of Mexico, Ecuador, and other parts. These ranged well back into fabled times of sunken Pacific continents, but they also surged forward into the end of the classical period; even Alexander the Great's lost fleet found a new role in a culturally fecundating voyage through the southern oceans to the western shores of the Americas.

Meanwhile, evidence of Phoenician, Egyptian, West African, Jewish, Roman, Celtic, and Viking contacts ranged from New England to Middle Eastern America in the North and down to Brazil in the South. Indications of a European or Eur-African presence in the centuries just before Columbus are not wanting.[159] The idea that the Americas were a virgin to the Old World before Columbus deflowered them is an anti-historical myth.

That there were many contacts seems clear. One has only to read Ameghino's survey of pre-Columbian encounters of the two regions, written a century ago as I mentioned earlier, to comprehend that, while he may have been naive, the contemporary scholar has been unreasonably skeptical. Moreover, much evidence has seen the light since his time.

As to the troublesome question concerning when these contacts took place, here we propose that the Americas have been in touch with the rest of the world throughout the history of mankind, except in the periods of great natural turbulence, with the contacts swelling in numbers whenever a few hundred years of technical development and cultural organization would occur. Whenever a catastrophe happened, which cut off peoples by splitting continental blocks, lifting mountains, creating great rivers, or interposing new climates between them, the isolated cultures developed very rapidly, requiring only a few centuries to exhibit different cultures, languages, and ways of life.

Let the editor of a recent collection of studies on trans-oceanic contacts summarize the situation for us:

> Clearly, the present status of our knowledge of American archeology does not allow us to attribute the origins of New World civilization to diffusion from the Old World with assurance. Equally, however, it does not demonstrate the independent origin of New World high culture. Just as the zero occurrence of artifacts originating in the Old World and found in America may be taken as a strong argument against the diffusionist explanation, so the early occurrence of a complex of Old World-like traits -- often very sophisticated -- in early levels

---

[159] Cyrus Gordon, *Before Columbus*, N. Y.: Crown, 1971 and *Riddles in History*. N. Y.: Crown, 1974.

of nuclear American civilization casts a strong reflection against the independent origins hypothesis.[160]

This points to a very early heartland culture; then came divergence and sporadic exchanges.

I would suggest, concerning said passage and the same anthology of studies, that we should be looking for several periods of transference of traits; in *Chaos and Creation* I suggest six of them. Artifacts and usages can then be assigned by ages and the outcomes tested (for their logic and verisimilitude). Basic social forms, early ceramics, boat design, the lodestone compass, the pyramid, Semitic, Celtic and Roman relics, and many other kinds of evidence exist with which to clarify the periods of intercourse.

To summarize, a hypothesis of ecumenical world culture in the earliest times, attaining quickly the Neolithic level, is supportable. Inventions require heavy motive power, both in the phase of mental gestation and of social adoption. The motive power must operate within and among individuals. Basic inventions came in rush following the gestalt of creation. They flowed from the psychology of the new human species, originally a small group. They were tied immediately to astral gods and figures and to animals as well; this identification lent memorial power to the inventions and authority to the thrust of their diffusion.

Acting in the name of their gods and totems gave authority to the imposition of practices. The same aggressiveness that ultimately eliminated the hominids also foisted upon them the basic inventions. Those who grasped the meanings of the human culture, or at least could practice it, survived. The aggressors possessed ideology, skills, and zeal. No species could stand against them.

In this manner an ecumenical or universal culture was quickly created and diffused among a variety of human racial types. Potentiated genes were diffused and came to the fore quickly in adapting to a changing world. Culture traits were imposed under the most stringent conditions. It was the greatest age of evangelism in the history of mankind. Within a thousand years of increasing natural terror, most basic skills would have been adapted from nature, developed, put into a framework of ideas and imprinted upon society.

---

[160] Riley, *op. cit.,* 457.

CULTURAL INTEGRATION

The Dogon people of the Upper Niger region of Africa have come to public attention recently.[161] Marcel Griaule's exposition of their secret lore has been presented by his collaborator, Germaine Dieterlen.[162] The Dogons have a rich astronomy. They know that the star system, Sirius, contains a bright star and also a dark, dwarf star, although it cannot be seen by the naked eye. Robert Temple studied exhaustively the sources of this knowledge and ventured the idea that astronauts from Sirius may have once have visited Earth and imparted this knowledge. Or else the dark star may have once exploded in a super-nova and was remembered. A third possibility is a one-time proximity of Sirius, which would imply a vastly accelerated expansive movement of the galaxy. Or a telescope. I incline towards the super-nova view.

The Dogon were probably survivors, with the ancient Egyptians, of the vast 'Triton' (Sahara) civilization that was destroyed about 6,000 years ago. In isolation, they have kept their knowledge accurately, obsessively, secretly. It took Griaule 16 years to hear the lore from them.

The Dogon culture shows clearly the fundamental law of cultural anthropology: All aspects of a culture are interconnected:

> The smallest everyday object may reveal a conscious reflection of a complex cosmogony... Thus for instance African techniques, so poor in appearance, like those of agriculture, weaving and smithing, have a rich, hidden content of significance... The sacrifice of a humble chicken, when accompanied by the necessary and effective ritual gestures, recalls in the thinking of those who have experienced it an understanding... of the origins and functioning of the universe.[163]

And we can quote the social theorist Cassirer also:

> If a man first directed his eyes to the heavens, it was not to satisfy a merely intellectual curiosity. What man really sought in the heavens was his own reflection and the order of his human universe. He felt that his world was bound by innumerable visible and invisible ties to the general order of the universe - and he tried to penetrate into this mysterious connection.[164]

---

[161] Robert Temple, *The Sirius Mystery*, London: Sidgwick and Jackson, 1976.

[162] Marcel Griaule and Germaine Dieterlen, *Le Renard Pâle*, Musée de l'Homme: Paris, 1965.

[163] *Ibid.*

[164] *Essay on Man*, New Haven: Yale U. Press, 1944, 48.

All the pieces of human culture resemble or hook on to each other. Social and body symbolism are international, for example, as Mary Douglas has shown, [165] also cosmogony and sex, diet and religion, and so on. Exceptions come from intrusions and novelties: these are rejected; but if lent power, persistence, and utility they will work themselves into the cousinship of culture traits. The discovery that this is so belongs to modern anthropology, to field workers such as Ruth Benedict, Margaret Mead, and Clyde Kluckhohn.[166] The discovery is in the air and an alert historian of science shares it. Thus Santillana writes : "As we follow the clues - stars, numbers, colors, plants, forms, verse, music, structures - a huge framework of connections is revealed at many levels. One is inside an echoing manifold where everything responds and everything has a place and a time assigned to it."[167]

Many studies pursue the First Law of Anthropology. Yet few ask why it should be. Why is a culture - Womburi, French Canadian, Hopi, Greek, or English - integrated?

"Because the human likes to be consistent." But why does he seek this consistency ?

"Because the human mind has to explain itself." Why so?

"Because all things *are* connected to the stars via the cosmos!" But in an industrial culture, millions of chickens are dispatched automatically without obvious connection to anything but the market for chickens.

Actually, all three theories hint at the best explanation. The human must be consistent in connecting all things, because, in the times following creation, culture burst forth spontaneously in all of its manifestations; all the objects of the world were not only to be seen, but also to be reflected upon, that is, to become objects of thought. Cultural consistency came before its rationalization. And each culture is of course culture-bound, viewing the world in its own way.

Since the days of creation must be obsessively remembered and repeated, as we shall see, they continue to force upon man their original togetherness. They supply the motive force for performing the greatest and the smallest tasks of society.

Then, too, since the burst of revelation and discovery was tied into the outbursts of the gods, all that is thought about becomes tied to the gods.

---

[165] *Natural Symbols: Explorations in Cosmology*, N. Y.: Pantheon, 1970.

[166] *Cf.* Kluckhohn. *Mirror for Man*, N. Y.: McGraw Hill, 1954.

[167] *Op. cit.*

Whereupon the human must realize this fact, confess it, and lend it importance, or else he will be guilty of blasphemy, ingratitude, and neglect of the gods. Hence he must excuse himself and his actions. Such is the explanation offered here of the First Law of Anthropology.

Every culture is integrated and coordinated within itself; this we know from the comparative study of existing cultures. All culture arose hologenetically, and diffused with the original homo schizo. But, in any event, they could not be radically different, because human nature sets limits on what a culture can do. We can hardly conceive of what might be different about cultures, because they are part of our very nature. Louis Wirth used to lecture that men differ in every way that it is possible to differ. If they do not differ otherwise, that is because it is impossible to do so. If it were possible, we would not know it. Further, there is no practice in any culture that lacks a homolog in every other culture.

The pattern and limits of culture began with and must follow the schizotypical nature of individual humans as they transact among themselves and with the world. Therefore, we can expect to trace the syndrome of schizotypicality through any given culture and all cultures taken together.

The recent insistence of some sociologists and ethologists upon the predetermination of human behavior does no more than make sense of the view that humans are culturally determined. Nature and nurture are inextricably bonded. One misleading view, which has flourished in many forms, is that culture is a thick varnish laid upon a brute to contain and rule him. To the contrary, humans are born to rule themselves and must spend their lives in trying to do so. They cannot ignore the problem of control. They must try promptly every conceivable means of doing so, whether this means reaching into their own nerves and muscles for the purpose or stretching outwards into the environment and then reimposing controls *via* a group and its culture.

Modern empiricists are often repelled by the mythologist who says that the ancients connected all with all. They cannot pursue the line of thought that connects everything - lines, crosses, comets, sceptres, circles, megaliths, and seemingly everything else - with a phallic symbol, for example. Or an eye with a comet, lightning bolt, an electric arc, a giant, a mountain, and so son. Anthropologists should make such connections as a matter of course; it is surprising when they do not.

There are two main reasons for granting that the earliest humans possessed a holoculture and thought in terms of it. One is the evidence itself, so voluminous that a thick book could be prepared of all the demonstrable, deliberate connections of the *membrum virilis* in tools, arts, stories, beliefs, and rites. But if the evidence is not overwhelmingly convincing, the quantavolutionary theory of early man should be. For the original humans -- and even the unconscious among the humans today -- thought in holistic terms. It is one of the lessons of logic, dutifully repeated in its textbooks, that 'analogy is not proof. ' But to the first homo sapiens schizotypus, and to humans of all times, analogy *must be* proof. The most marvelous sense of power, intellectually and behaviorally, comes from the association of the tiniest events and observations with the nature and conduct of the great universe.

Here the anthropologists, the mythologists, the pre-historians do agree. All things are tied together: a sacred universal bond exists among all things. One may imagine that millions of hours went into both fantastic and carefully considered leaps in order to form all sights, sounds, and experiences into a meaningful whole.

The ability and need to see all in all is fundamental to the newly created human. The scientifically and technically useful ability to concentrate upon only a single special aspect of a thing derives from the obsessive compulsion to repeat.

The two needs spring quickly from the urge to control. Fearfully and paranoically, the humans saw in everything the thing that would threaten (or, ambivalently, save) them. Fearfully and obsessively, humans had to rehearse and redo what they had experienced, keeping everything the same and in order.

CHAPTER VI

SCHIZOID INSTITUTIONS

Totem and taboo organize and report 'right' and 'wrong' for the people of a culture. They control one's selves by setting up a bank of animated displacements, publicly symbolized, and preventing one's selves from disturbing the assemblage. It would seem to be a normal way for homo schizo to behave. It does not matter that the terms are reserved for 'savages; ' civilized cultures can and do employ the totem and taboo. Most of this chapter, once it moves from the opening theme, plays upon their variations.

Totems and taboos are convenient ways of repeating and organizing obsessions. They are group elaborations of the schizophrenia of original humans. Both are found in all cultures and in varying degrees of weight. In large-scale cultures they are part of religion and bureaucracy.

Taboos are sacred prohibitions, whether received directly or indirectly from divine authority. The 'Ten Commandments' include taboos. The name of Yahweh was taboo. At one time it might be pronounced only once a year. Violation of taboos is commonly supposed to have fatal results. Yahweh frequently concludes his injunctions with the phrase "... lest you die."

The totem, more strictly, is a symbolic identification of a human group with an animal or plant, which represents a divine force. Because animals (the owl, for instance) and plants (the sacred oak) were tangible, near at hand, and well-known, they could readily be fitted into the scheme of delusions; a communication system, largely imaginary, is set up between the life-form and its human patron.

In joining with a totem, a human group acquired a talisman and group representative. The totem life-form operated in the sky and on earth to the presumed over-all benefit of its sponsors. Once the sun (earth) rotated too fast; the great rabbit, said some American Indians of the Great Plains, lassoed the sun and halted it, not releasing it until it promised to go slower (perhaps the rabbit was a cometary image.) But a totem also imposed limitations upon behavior by means of taboos, rites, and penalties.

Totemism came to be a set of specialized practices with regard to a species or even a particular animal or plant. It arose with the help of certain celestial behaviors that were for various reasons interpreted as animate behaviors within the celestial environments. The important illusory behaviors of the animation in the sky are carried down to Earth and cemented by analogy to the organism's earthly behavior. Thenceforth a set of attitudes to the life manifestations are produced that give birth to totemestic practices. As the human draws apart from the 'lower forms of life, ' the totem and the taboo dissolve into sublimations.

A totem provides a complete schizotypical system: the injection of divinity into an animal denotes a cognitive disorder, a hallucination, a misplaced metaphor. The exclusiveness of the totem and its group towards other totem groups in its associated taboos reflects the schizoid aversiveness to others; the worshiping and cannibal sacrifice (sometimes) of the totem animal emerges from ambivalence; the numerous rituals and rules connected with the totem convey compulsive obsessiveness; and the secret and enduring aspects of the totem group's practices, going back to the totemic primal incident, display catatonism.

Enrico Garzilli writes of Faulkner, Joyce, Pirandello, and Gide in their searches for "the real self," notable names, to be sure, pursuing at the pinnacles of literature the primordial search for oneself within the polyego.[168] He explains that the word is the self; becoming human is to become a word. Hence the importance of such ancient expressions as begin the Gospel of John: "In the beginning was Word; and Word suffused God; and God was Word." (My rendering.) We sense here the power and control exercised in the first naming of something and agreeing upon it with others. We should understand, too, here, that "Word" is "Logos" or "the enlightened life of mind."

---

[168] *Circles without Center: Paths to the Discovery and Creation of Self in Modern Literature,* Cambridge, Mass: Harvard U. Press, 1972.

SPEECH AND LANGUAGE

C. Levi-Strauss is of the opinion that "language was born all at once," thus supporting our position of hologenesis. He goes on to say that "whatever the moment and the circumstances of its appearing in the range of animal life, language has necessarily appeared all at once. Things cannot have begun to signify gradually. After a transformation the study of which has no relevance in the field of social sciences, but only in biology or psychology, a change has taken place, from a stage where nothing had meaning to a stage where everything had."[169]

This is a surprising use of the word "relevance." Once we have understood what was happening biologically and psychologically, we comprehend what was happening socially. A quantavolution introducing language must concurrently involve a grasping for logic, for control over memory, and for the social consensus on meanings from which culture sprouts. We have already spoken of what was happening biologically and psychologically: the hominid's brain was beset by delays in instinctive reactions, building special sub-centers, and displacing throughout himself and the world outside. The internal code of language was springing up and erupting here and there into public language.

According the Edward Sapir, too, language was formally complete from the beginning and existed from the beginning of man. H. Kalmus claims an "explosive" origin of speech, too, but then limits the speed to "hundreds of generations,"[170] a retreat to appease the millions of years of mankind awaiting fulfillment. Speech did not occur word by word, grammatical form by grammatical form, over millions of years of humanization. It probably sprang up in a mixture of counting, signs, and ejaculations. Counting has been connected (through Lord Raglan's *How Came Civilization?*) by Seidenberg[171] with rituals, which fits the model of homo schizo well. "Counting was invented in a civilized center, in elaboration of the creation ritual, as a means of calling participants in ritual onto the ritual scene, once and only once and then diffused." Seidenberg explains that all people had religious numberings and taboos on certain kinds of counting. It is frequently imagined to be theft when one's name is counted. Today, a homologous paranoia underlies the hostility of many persons to the

---

[169] Introduction to M. Mauss, *Sociologie et Anthropologie*, Paris: Presses Universitaires, 1968.

[170] In Frank Smith and G. A. Miller, eds., *The Genesis of Language*, Cambridge, Mass., MIT Press, 1966, 282-8.

[171] "The Ritual Origin of Counting," 2 *Arch. for Hist. of Exact Sci.* 1, Berlin: Springer Verlag, 1962, 1-40.

computer, which seems to steal one's name, carry one's number, and manipulate these and hence oneself.

A person is only created when named or announced and the creative word may have been the creative number. Marshack would seem to be moving along a similar path, with stress upon arithmetic and calendarizing (the catastrophized need to watch the skies for regularities that are hoped for, and irregularities that one must prepare for).[172] He is also locating ever earlier symbolic forms.

Some anthropologists are proving that the chimpanzee can learn to understand words and sentences. The point of exhaustion is reached after several dozen of them are learned. If the chimpanzee has not learned to speak in its supposed eight or more million years of existence or whatever its age as a species, it is unlikely to begin now. On the other hand, if the chimpanzee had just recently been mutated, the effort might be worthwhile.

The human seems better equipped to move his tongue than the chimpanzee, but it is not the primate's tongue that prevents speech. "Basic English," a shortened selection of words for communicating in English, does well with 750 words from a possible quarter of a million. (Its problem lies in the constructions; the format or program of a language would be critical to a world tongue, and cannot be simply imperialistic.) A number of gods have as many names as would be needed to constitute a language, hundreds for every major god.

Many vertebrates and insects could manage 500 distinct sound-combinations; 9 distinct sounds might be permuted about 2^9 or 512 ways. Since words have several meanings, depending upon their context, a great many more than 512 'words' are possible. When these thousands of words are combined, many thousands of messages are possible, enough to make a lexicographer out of a sparrow.

In order to speak, an animal has to be "intelligent." This means that it must possess a sense of being an individual, a will to words, the things to refer them to, a capacity for time and recall, and an obsession for reiteration.

There is no speech center in the human brain; a large cortical area controls speech and is placed in either the left hemisphere (for the right-handed) or the right. This would suggest not only that speech is recent and non-organic in structure, but also that the will to speak is an inner necessity connected with instinctual blockage between the left and right hemispheres, and slowdowns in message transmission in other newly grown parts of the brain.

---

[172] Alexander Marshack, *The Roots of Civilization, The Cognitive Beginnings of Man's First Art Symbol and Notation,* N. Y.: McGraw Hill, 1972.

Man did not get so clever that he began to talk. He was originally so frightened that he began to ejaculate names, and to call them out obsessively, then to use them on like occasion (to compare, in effect), to admonish, to pray, and command. To his surprise, he could find others who might understand, at first perhaps only a twin, then their offspring. Nouns came first, wrote G. Vico, one of the earliest modern etymologists. And he definitely connected the earliest speech with the worship of the gods.

Following the ejaculative phase, which may have occupied only a few years, language probably entered upon a liturgical phase. Heavily depending upon exclamation, it moved to detailing situations and meanings. It undertook to express what had happened (to call the roll of disasters, so to speak), to exorcize the causes of the events, and to cover them up, making sounds of appeasement or evasion.

Much public or formal language, like liturgy, has been formal and compulsory from the beginning. It is still so, obviously in mega-societies but also in tribal societies. Maurice Bloch speaks of the deliberate and enforced impoverishment of language in traditional oratory. The language acts to control the speaker.[173] He cannot go beyond prescribed forms of speaking. Hence public speech is understandable only in the context of ritual, as Malinowski said, not by virtue solely of knowing its lexical units. The rhetoric cannot become revolutionary.

GRAPHICS

Speech came promptly, but writing was not developed well until civilizations had poetry, art, religions, and social systems. A possible reason for this may also be supportive of our theory of language. It is logical that ' as speech is to the mouth and ear, writing is to the hand and eye. No one doubts that earliest man (or latest hominid) was as digitally adept as he was orally proficient. However, gestures, grimaces, and context could let the eye help the speaking process along.

But the hand and eye could not, like symbols, accomplish internal symbolizing or speech, which is probably what was occurring in the new creature to help him coordinate his several selves and their displacements in the outer world. That is, public speech was the extrusions of inner speech, like the small portion of the iceberg that floats above water.

---

[173] *Political Language and Oratory in Traditional Society*, London: Academic Press, 1975; *cf.* Charles Morris, Signs, Language, and Behavior, N. Y.: Prentice Hall, 1946.

Some people with complex languages do not write even today. Art of course takes the place of writing in respect to many messages from one's ancestors. A totem pole can take the place of much written history, depending upon the kind of history wanted. There is a clue here: a large society and an official class need explicit messages and records.

Until these criteria come into play, art can successfully block writing, somewhat as television blocks literacy. Art can say so much that, by comparison, the breaking down of pictures and symbols into writing may appear to be a meaningless and barren enterprise. Further, it may seem to be sacrilegious to openly admit that words are interchangeable tools. Hence writing was originally a holy profession, as in the Egyptian bureaucratic empire. It was carried over into government: "the needs of a centralized administration were a far greater impetus to the development of writing, among the Sumerians (cuneiform) as in Crete, than intellectual and spiritual needs."[174] Earliest tablets speak mostly of rations and personnel in the palaces.

But, in maritime cultures, such as the Phoenician, the pragmatic value of messages finally broke the sacred grip. Words (orally spoken) had departed so far from their origins and symbols from art, that they might be used casually in practical affairs. The alphabet was invented out of numbers, phonetics, and calendars by people who were on the move, as in boats.[175]

The invention of writing was an effective grasping for control of memory, behavior, and pragmatics. It delivered also a severe blow to the imagination; it caused massive disenchantment. It placed credit for works effectively upon the culture. No longer could one be taught by the gods, through subtle or at least mysterious parental and social transmission or from the depths of one's being, from inner springs. Besides memorization, one had exactness, repetition, a third party, an objectivity, a beginning of coolness and remoteness.

PRIMORDIAL LANGUAGE

Man spoke one tongue to begin with. As he diffused from his proto-patria, his speech had reason both to change and to remain the same. If there can be found a basic set of sounds and words that is common to all of mankind today, then one would have an original language, a proof of cultural hologenesis, and an indication of the recency of human origin.

---

[174] M. I. Finley, *Early Greece*, London: Chatto and Windus, 1970.

[175] Cyrus Gordon, *Before Columbus*, 103ff.

Searches for the first language have been modestly rewarding, enough so to justify a greater expenditure of time and resources, especially for computerized manipulation of data. R. Fester has proposed that "there is an original vocabulary of six archetypes common to all of humanity which still today comprises the basic of every language and which at the same time provides a clearly recognizable link between all languages." The root-words of 'Pangean, ' as we might call the tongue, would be BA, KALL, TAL, OS, ACQ, and TAG. "From the moment when the genus *homo* left the family of lower animals, and thanks to his upright stance, both hands and *senses* could serve him more freely than before, the *vox humana* shared his further evolution to the Man of today." [176] We should, of course, disregard the makeshift ladder that Fester has thrown up here to arrive at human voicing. The words are prominent today in geography: "Indo-European, Mongolian, Phoenician, African and Ancient American geography was decidedly using the same original words."

Fester claims to have discovered that in many languages, the syllable BA pertains to human relations and subsistence; KALL appears connected with the idea of concavity and the females womb; TAL refers to clefts, to the ground, to females; OS to thresholds; ACQ to water; and TAG to height, gods, erect humans. To Malcolm Lowery, who has kindly supplied me with his translated materials, the progression by which the words related by Fester to the roots are said to drift in space and among cultures is not intelligible.

J. P. Cohane also proposed a set of root words, independently and without awareness of Fester's book.[177] These key words, he believes, were strongly religious in their original associations. Like Fester, he finds his examples to be most copious in geography. His words are also six in number, although others of equal importance seem to be present in his narrative. They are Oc (or Og) as in Okeanos, Kronos, Moloch, and an ancient Irish god, Oc; Hawwah, as in Aloha, Yahweh, acqua, earth; mana; ash/ az; tema, as in Thames, Tiamat, Athena; and Eber/ abar, as in Berber, Hibernia, Calabria, Abruzzi, Hebrew, Ares, Mars.

Scholars of linguistics seem disinclined to undertake the risky task of reconstructing the prototype language. Whorf spoke of "the story of man's linguistic development -- of the long evolution of thousands of very different systems of discerning, selecting, organizing, and operating with relationships. Of the early stages of this evolutionary process, we know

---

[176] *Sprache der Eiszeit: Die Archetypen der Vox Humana*, Berlin: Herbig, 1962, 31, 6.

[177] Cohane, *The Key*, N. Y.: Crown, 1969, is directly comparably with Fester.

nothing."[178] We can, he said, only survey the results of this evolution as they exist today. Still, Whorf was an early enthusiast for trying to trace the original ecumenical speech.

Generally, the linguistic establishment has beaten back the numerous efforts to demonstrate speech affinities, regarding them as *prima facie* absurd. Such connections would be Gaelic with Algonkin, Chiapenec with Hebrew, Othomi with Chinese, Choctow with Ural-Altaic, these being Amerindian connections. The diffusionists have fared better in proposing Old World connections: Hamites with Semites; Sumerians with Magyars; Late Minoan with Greek; Egyptian with Hurrian; Etruscan with Lemnian; Berber with Basque, etc. Justus Greenberg says that the 750 indigenous languages of Africa were originally four families, and these were originally one, and possibly related to Hamitic, says Gilbert Davidowitz. Encouraged by the theory of hologenesis of culture, I would conclude that the search for the ultimate ecumenical Pangean language will not be in vain.

GROUP VS. INDIVIDUAL

Humans of the proto-age had immediately the problem of constituting themselves deliberately into a group. The psychology of the hominid band was gone. In its place was the fearful, distracted, individuated - even multividuated - person. He must belong, yet not belong, at the same time. The favorite topic of political philosophers and economists - the individual against society - took shape.

The bond between individual and collective psychology is tight. It is both genetic and adaptive. It is fully determined. It is unbreakable. Evidence of these statements gushes from history and anthropology on the one side and from many psychological schools on the other. Just as the brain can reach to the toe to express itself physiologically, it can reach to the stars to express itself psychologically. Where it happens to reach is a cultural affair.

Just as the human is a coordinated poly-ego, so a culture, and for that matter any group, is a mega-poly-ego, that typically selects a dominating ego-pattern as its design for the behavior of its members. A special concept of organization is required to grasp that organized behavior that is an extension of patterned mind-behaviors. The genesis of external organization is in the mind(s) of individuals and their groups.

One way of expressing the holism of personal human conduct is that "private motives are displaced onto public objects." Thus, a person suffering inferiority and weakness in personal life finds superiority and strength in

---

[178] B. L. Whorf, *Language, Thought and Reality*, Cambridge Mass: M. I. T. Press, 1956, 84.

political activism; Harold Lasswell, following Alfred Adler, has expounded and documented this thesis.[179]

I do not limit our theory to this view or language. All men, given their brainwork problems, must feel weak. All men seek power according to their own private and cultural prescription. The distinction between private (individual) and public (social, cultural) is most usefully applied during special investigations in politics and law. The human bonding is without innate distinction. The human acts in a merged internal and external context. A fond pat on the hand can stop a pain in the toe; a political victory can let a man digest a thick steak, as I once observed in a study of Huey "Kingfish" Long of Louisiana.

There is no end to the process of 'private-public' interaction from conception to death. That means also private-cultural. The individual and the group march along, side by side, from the dawn of mankind. Both society and the individual are schizoid in origins, structure, and functions. Their behavior and forms are not always congruent; the symptomology is varied. Then it is that deviance (medical schizophrenia) is defined. The individuals seek to evade the society or change its laws; the society seeks to make the individuals conform; else it treats them for mental illness or jails them on account of their menacing or destructive conduct.

The process will go on as long as human nature retains the form which it assumed in the days of creation. There are perhaps some non-schizoid culturally created humans, who have evaded hybridization with the schizoid, the fate of most hominids. Even if there were none at all, the idea of their existence should be retained for heuristic and theoretical purposes. They would be well-trained primates, although not discernible as such. The schizoids, and especially certain schizophrenes, are religiously and politically dominant. With their obsessions, suspicious hyperawareness, penchant for symbolism, and their megalomania they control the world. That is, they try to control it; but the world is, by their own definition, uncontrollable. Homo sapiens schizotypus defines 'control, ' and is insatiably anxious for control.

Human action moved back and forth along an axis of tension between the individual and the collective or social. Self-awareness was an inescapably individualist phenomenon. Never after creation could the sense of the self be exterminated. Never thereafter, then, could the collectivity perpetually and wholly dominate the individual soul. Incessant, forcible, and imaginative attempts to do so over all of history were foredoomed to fail and still are. The split self, a source of the greatest terror, could not permit its unification by the collectivity, even though the collective achieved its great resilient

---

[179] *Power and Personality*, N. Y.: W. W. Norton, 1948.

strength from its guarantees to the individual that it would assuage, diminish and even cure the terror of the split. There was no returning to the mammal.

So loyalty began, built upon intrinsic disobedience. And so began authority. The story of Job, in the Bible, represents the individual trying with all of his might to subject himself to the will of Yahweh. Dreadful catastrophe, initiated by Yahweh, abetted by the Devil and by hostile humans, crushes his life-values: his loved ones, his possessions, his power, his respect, and his health. An exception stands for the sixth value, knowledge, that is not removed but is the focus of the divine assault upon Job. If only he could be mentally broken into a numbness, stupefied, then he could be defeated. He would not then respond to God.

The very failure of this last form of degradation of self is both a triumph and a negation of Yahweh. That is, all must stop short of the ultimate disaster, which would effectively wipe out creation. On the other hand, once stopped short of self-effacement, the campaign of Yahweh and the Devil is lost and the human being is restored. Job is left the victor on the scene of battle. All of his values and achievements are indeed restored. The story of Job is told as a lesson in humility; actually, it is a lesson in human arrogance: the will to control God.

Job's story might be set, symbolically, at the end (ca 4000 B. C.) of the age of Elohim-Saturn. It is before the flood of Noah. By then, human ideation was as complete as it was to be until the Greek skeptics, unless some civilization, of which no trace remains, had operated with a secular ideology. Technology had arrived at a level hardly exceeded until 350 years ago. At Catal Hüjük, in present-day Turkey (6,000 B. C. ?), "orderliness and planning prevail everywhere; in the size of brick, the standard plan of houses and shrines, the heights of panels, doorways, hearths and ovens and to a great extent in the size of rooms."[180]

During the age following Saturn, which may be called the age of Jupiter (Zeus, Horus, Yahweh, Marduk), the list of secondary institutions and inventions becomes long. Large scale organization or centralization developed. Millions of people were aggregated and ruled by agents and delegations of authority. Kingship; priestly, military, and official classes; record-keeping; and extensive physical properties were common. Increased domestication, breeding, and herding of varied animal species reflected a projection of human organization into the animal kingdom. Large-scale agriculture is also to be viewed in the context of an administrative organization of plants and human caretakers.

---

[180] W. A. Fairservis, Jr., *The Threshold of Civilization*, N. Y.: Scribner's 1975, 143.

PSYCHOLOGY OF ORGANIZATION

Basically, given the domineering schizoid prototype, social behavior (including language, religion, governance, art, etc.) contains varying elements of obsessiveness, catatonism, orgiasm and sublimation. The fears of the self, of the gods, and of loss of control lead to the eternally 'shell-shocked' behavior of returning to the original traumas and repeating them, both to punish oneself and to avoid punishment by others. Deviation is tabooed, except as it finds expression in momentary orgiasm and sublimation.

The only way in which language and all other inventions of customs can be developed and organized happens to be schizotypical: undeviating insistence upon repetition, the compulsion to repeat, the slavish adherence to memory and tradition, liturgies. Bleuler reports patients who will play the same musical trill or chord a thousand times and, like the esteemed citizen of the regimenting modern state, Bleuler's patient, obsessed with command automism, will mechanically obey any outside order, will imitate others slavishly, will repeat everything he hears, and, despite a lack of feelings, do all of these things impulsively or as if compelled.

This is an effective human response to a loss of instinct and the great need for new forms of control over the self and others. Organization, even as we see it today in great bureaucracies, highly rationalized, is a catatonic gripping for a non-changing world: 'If I remain perfectly still, I will escape observation, I will not be punished, and the world itself will stand still in emulation of me.' Members of a Judaic sect freeze in whatever activity they may be engaged when the Sabbath falls and do not move until the Sabbath ends.

In its conception and supposed functioning, a typical modern bureaucracy is a marvel of deductive science.[181] It is hierarchy of power and control from top to bottom, with a division of tasks from broader to more narrow scope, down to the individual worker. It is regarded as a highly rational way of accomplishing large collective tasks. Yet this administrative grandeur is only the recognizable descendant of the first efforts of homo schizo to organize work, something he did half-aware but naturally. For the principle has been the same from then to now: an obsession upon a displaced target (god, a village plan, a hunt, agriculture) and an effusion of severe discipline, compulsively exercised and rationalized. Man has had to work in this way. The awareness of the principle, its statement in science

---

[181] A. de Grazia, "The Science and Values of Administration" *Admin. Sci. Q.* (Dec. 1960) 363-98; (March 1961) 558-83.

and law, and deductionism as scientific method all trail after its spontaneous generation.

In early organizations, the compulsion to reiterate was applied to external control and organization as it had originally been employed for self-control and the ordering of smaller groups. Authority was supplemented by deductive principle. Deductionism is the idea that from a general prescription may be derived specific prescriptions. That is, a statement, that all must be put in strict order, is followed by an enforcement system to ensure that no exceptions to or deviations from the order occur in individual cases.

Deduction is consistent with the association of different kinds of displacements and the compulsion to reiterate. It permits free play to authority to expand its scope of activity and its human domain. It leads to all avenues of life. It externalizes the subjective, by providing security, letting the inner self relax, and divesting the self from its preoccupations with itselves into 'objective' external occupations. It relieves the smaller social organizations of their involuted and intricate rites and rules, moving them out upon the larger stage of a kingdom.

Constructions of many types became possible. Monuments, settlements, populations, armies, and record-keeping all grew in size. A bureaucratic (usually theocratic) state might be discerned, successful in its aggrandizement of human activities, and containing within its larger order the orgiastic practices of religion and warfare, the sublimatory development of the arts and crafts, and the negativism and retardation always imminent in human populations.

Bureaucratic states might collapse from natural disaster, or from competing states, or even from long-term demoralization. Deductionism is rigid and restrictive. It puts constraints upon ambitions, social differences, and new experiences (orgiastically impelled). It is prey to apathy.

Nonetheless such social forms as the bureaucratic kingdom must be called a civilization. The surrounding and preceding forms might also be called civilizations. When, then, did civilizations begin? Civilization is premised as some condition beyond humanization. The human could not elect civilization; he was driven to it by his fundamental character; what was needed was a respite from catastrophe and a space of a few centuries.

Civilization marked no qualitative change in the human character. It is an enduring, well-grounded way of life for a large number of persons containing elaborated and sublimated second-order effects of humanization. If more severe strictures are put upon the term, no significant benefit in logic or theory accrues. Writing is civilized, but provokes no great change in human character or ideation. Deducing commandments from a generalized authority is not exclusively a civilized practice. Peacefulness is not exclusively a trait of civilization. If it were not for the catatonic motif that freezes many cultures at a first-order stage or in a 'fallen' stage, the word 'civilization' could be logically applied to all human organization.

The catatonic response to disaster may be presumed to account for a number of 'primitive' or 'retrograde' peoples and subgroups of larger populations, such that the elaboration which is the hallmark of civilization does not proceed. This catatonism is negative and refuses change. It fights the battle for world control within the person and the small clan or tribe. Its overburden of constraints, rejections, and taboos miniaturizes and trivializes. The externalized, exo-tribal culture is actually abandoned and condemned, leaving the members of the group motionless, aghast, face to face with awful eternal threat.

MEGALITHS AND MEGALINES

People built megaliths around the world, probably beginning six thousand years ago. A megalith is a worked or cut stone that weighs, say, over 10 tons, which alone or in conjunction with other stones mediates religious sentiments among the group and with the gods. The stones stand for ancestors, gods, holy circles, sacrificial altars, astronomical pointers, centers of convocation, and tombs. The efforts required to erect them demonstrate both strenuous collaborative discipline and fervid emotions. They are large to demonstrate the peak of divine fealty of which the group is capable and to stand firm against the elemental rages of nature.

That they are often isolated from their quarries or sources, have been reconstructed again and again, have been abandoned by or remain from a disappeared culture, and are fallen, split, and cracked indicate that the fears of their builders were well-founded. The builders were dispersed or annihilated. When recently the megaliths were rediscovered and studied, they were considered mistakenly to reflect a peak level of technology of their builders. Actually, many of them may represent the work of marginal surviving elements from civilizations that peaked at higher technical levels but whose centers were eradicated.

The Olmecs of the Mexican lowlands used basalt quarried from eighty miles to the North to build their monumental sculptures. Single stela and single heads weigh from forty to fifty tons. "The scale of the operation required dwarfs that of Stonehenge and speaks for an authority of great power at La Venta, backed by potent sanctions."[182]

Gerald S. Hawkins examined the famous Nazca ground tattoo of lines for stars, planets, sun and moon alignments and found none. "The line complex was not built to point to the sun, moon, stars, or planets. Astronomically speaking, the system is random."[183] But, if the Nazca lines are very old, we could expect them to be nonsensical by current retro-reckoning in astronomy, for there is evidence that the Earth has tilted during human times. Thaddeus M. Cowan, a psychologist and archaeoastronomer, writes that Old World proto-astronomers

> ... were primarily concerned with significant solar and lunar events as they appeared on the horizon.... Indian lore suggests a variety of ways the stars can be regarded (individual stars, groups of individual stars, patterns). Similarly, the mounds might be seen as following the same course (conicals, chains, effigies).[184]

That is, to the Amerindian, the mounds and stonework were templates of the constellations and sky events, therefore measured large (though subjectively) and mostly not even fully visible from the ground and to the workers. Probably this is behind the Nazca drawings too, with their spectacular drawings of dragons and a large bird.

Since the megaliths were all constructed before the Earth and sun had achieved their present orientations, none of them preserve their original orientations. At best they point roughly towards some anniversary position of the sun, moon or stars, such as the spring equinox. They were not the best instruments, either, of their times and culture. Whatever the orientation and cyclical repetition that they counted and measured, a simpler, more manageable, and finer measure was within the builders' capabilities.

The megaliths were religious. The Arc de Triomphe in Paris was built to celebrate a concatenation of heroic and historic deeds, not to be simply a convenient traffic divider. It was oriented to the sunset of the victorious December 2nd Battle of Austerlitz, also Napoleon Bonaparte's Coronation Day. So too the megaliths were erected, not to count months or praise the

---

[182] Michael Coe, "Mesoamerican Astronomy," in Aveni ed., *op. cit.*, 89.

[183] Gerald S. Hawkins, "Astroarchaeology: The Unwritten Evidence," in Aveni, ed., *op. cit.*, 89.

[184] "Effigy Mounds and Stellar Representation: A Comparison of Old World and New World Alignment Schemes," in Aveni, ed. *cit* 218-35, 222-3.

dead, but to commemorate the past, to celebrate survivorship, and to control destiny.

ORGANIZATION AND CONTROL

The first task of the split-self was to recollect itself and gain control of itself and others. How it did so became the paradigm of governance ever thereafter. The workable mechanism incorporated the overflowing stored fear, the gods associated with its origins and still operative in the sky and on earth, and the four patterns of behavior, all of which could be brought to bear upon the problem of self-control and the control of others, lending the person a tolerable balance of mind and behavior while identifying with and yet subverting the gods and accomplishing the pragmatic functions of existence in a much more developed and technical way.

All of this appears rational and fully intentional only in retrospect. Most of it occurred as the reaction and response of a new species of being to continued applications of great internal and external stress. The bearers of the new human culture were not all members of the new humanity. Whatever the combinations of mutation and potentiation, schizotypical leaders were present from the beginning, who, in order to adapt themselves to the new life, had to seek the adaptation of the others. They possessed all the tools of leadership that have ever since been possessed, namely symbols, ideology, force and goods, in the same order of importance, all tied together in the drive to control and organize the environment according to a teleological, if delusional, form.

The predominance of the delusional, aggressive, symbolist character in governance began then and continued ever after. Human food production, and the useful arts and crafts could not move forward without qualities of leadership removed from the actual specialization of tasks, no matter how important and vital they seemed in themselves. Control and power in the self and in the group were their preconditions. Therefore priests were the governors, and hunters, planters, and workers the governed. In a later elaboration, god-kings assured the society a personalized succession from the gods under covenants and constitutions; these the priests contrived to tie human governance to the order and disorder of the skies. The universal presence of generalized, rather than specialized, leadership is knotted to the principle of the total cohesion of culture, earlier described. Since the culture is holistic, so must the culture's leadership be holistic.

REPUBLIC AND MONARCHY

The basic political institutions are but two, the republic and the monarchy. All human organization resolves into a combination of these. The republican is of hominid origin and was the logical first form of human organization. The monarchies originate from the catastrophes following creation and the relentless evolution of homo sapiens schizotypus.

By the time of the first extant historical records, ca. 5200 B. P., past ages and past god-kings had reigned and retired. Now there came, in what is conventionally regarded as the first dynasty of Egypt, a consolidation and a worship of the sky-god Horus (Jupiter) and the Pharaoh as divine king. But this unifier of Egypt built his rule upon a congeries of small kingdoms each with its own divinities and cosmogonies. It is said that in the control of the Nile waters (perhaps after the Saturnian flood) lay the appeal for the unification of Egypt. More important as a cause was the aggressive force unleashed in the aftermath of disaster. Just as the divinity of a king is proven by his internal absolute power, and that proof is rendered necessary by a natural (divine) destruction of the previous power of the prior dynasty and rule, so is his divinity proven by his ability to master foreign societies, in emulation of the universal omnipotence of the king in the sky. The Pharaoh incarnated Horus.

The sequence of rulership tended to proceed from disaster to survivorship to monarchy to republic and then through the same sequence repeatedly, over cycles of varying duration, until the cycle became a self-fulfilling prophecy, or Plato's tyranny, aristocracy, democracy and so back to tyranny, aristocracy, democracy and so back to tyranny, a law of politics. This could be rationalized, without the recollection of primeval catastrophe, as 'the way man's mind worked' and 'how societies changed.' Even to this day, the cycle tends to occur. The chaos that typically ends democracy is laid to libertinism, rather than to the subconscious primordial feelings excited by a rule of liberty.

AUTHORITY

The end of democracy, that is, comes not from what happens but from an increasing feeling that 'man is getting away with too much, ' and that the gods will respond by devastating man. So a tyrant arises, plays god, restores order, and people hope that their expiation and sacrifices of their liberties will be punishment enough. Usually the occasion for the crisis of the regime and the revolution of the government is something resembling a catastrophe: a natural disaster such as a drought, a hurricane, crop failure, economic depression, or a crushing defeat by a hostile army. The function of authority is to support the structure of the human mind that was erected upon the dire events that brought the human mind into being. Authority is the formula that encompasses the three control-needs -- the control of the self, which is paramount; the control of the gods; and the control of the environment, including other people.

The world may be believed to consist of an objective reality but that objective reality is a product of an uncertain mind. The objectiveness of reality consists of a mind that perceives itself and therefore perceives the need to define reality, plus an agreement of many minds that reality is as it is. Both come from the shared structure and discipline of the newly create humans.

But the mind is uncertain of this absolute reality and human society is an endless struggle to set up and maintain this reality against the indecisiveness of human instinct and the discrepancies of perspective, both genetically and experientially caused.

The bonding consists of a) the projection and identification of the mind with all of these together, b) the obsession (repetition compulsion) as a glue of the binding, c) the deductive principle as the method of moving through time and space and dealing with all three components while moving, d) the lesser principles of catatonism, orgiasm, and pragmatics that are intrinsically incapable of ungluing the binding formula of authority, unless and until the mind is destroyed. That is, not only was chaos the primeval cloud-world, formless and kaleidoscopic, but also chaos was his non-recollectable existence to which he could not return, and feared, and therefore would not wish to go back to.

COVENANT AND CONTRACT

Men have always cherished the hope that the gods would cease to torment them. One of the most brilliant inventions to bring this about was the 'covenant' of the lord. The gods would promise to perform certain tasks and refrain from harming people provided that the people would worship them properly and behave in certain ways as well. In the Bible, Elohim and Yahweh introduce at least seven covenants. One is with Adam and Eve, another with Noah, others with Abraham, Isaac, Jacob, Moses and David. Several of these followed upon natural disasters. A new constitution, so to speak, was handed down from the throne.

But note how the route from catastrophe to theocracy to monarchy to individualism is pursued. First there is chaos: no promises are binding.

Friedrich Nietzsche, predecessor to Freud in the discovery of the 'unconscious,' writes in the *Genealogy of Morals* that the human was originally simply a fickle animal. Somehow the creature had to be severely chastised in order to give it a memory. For the keeping of promises was the basic condition of humanization and civilization. Surprisingly, Nietzsche does not take the leap to catastrophism, inasmuch as he thought that some immense event must have happened to cause mankind to acquire a memory. He says that he cannot think of anything more severe than the punishment that would be dealt out to persons who did not keep a bargain in early tribal commerce.

Tribal commercial promises, like many another cultural trait, are traced by the quantavolutionist to the by-product, the fall-out, from the great reality of chaos and creations: 'You made us; you have destroyed us; do not do so again; we must believe you will not; we have this specific assurance from you, your self-binding covenant; promises must be kept (we hope) and therefore we shall kill any among us who violate your covenant; further we will punish anyone who violates any promise that he makes, and go to war over broken promises; so all contracts shall be sacred in your name.' Thus I would imagine the genesis of contracts.

Then as egalitarianism progressed, the laws came to regard a great many contracts as made between equals, rather than handed down as in the beginning. So the sanctity of contracts, which the American courts for a long time expounded with holy fervor, goes back to the times of reaction and destruction, to fear, and to the magical coupling of the sacred and the profane so as to reinforce the sacred. The pragmatic element of the contract is of course great; there is no gainsaying Nietzsche there; but the catastrophized essence is there too.

As in other areas, the catastrophized behavior works itself out in a highly sublimated and indirect form. The 'rationality' of the 'contract law' is on the one side. The famed penchant of Jews for 'arguing with El,' 'legalism,' and 'bargaining' derived in part from their catastrophized anxiety over whether a new covenant would be pending and what the words of the last covenant really meant.

On the other side, the aboriginal idea runs rampant. The Zealots of the first century after Christ are a case in point. They believed a terrible calamity would soon overcome the world and wipe out all but a few good Jews. Then a new covenant would have to be confirmed by Yahweh. St. Paul extended the time, and modified this idea into a belief in the Judgement Day. But the early 'millennialist' sects are imitated from time to time today.

And a close homology is afforded by the 'Cargo Cult' sects of the Pacific Islands; there, property and promises are dispensed with, in a fatal pause or age-breaking, during which a people awaits the coming of a great ship or (now) airplane carrying the goods of life promised by a sacred ancestor.

We noted how the Greeks handled the problem of promises. Their gods were full of not-so-valid promises, possibly because they were so full of obligations and interconnections. Whence their gods were deemed fickle, not at all like Yahweh. But, recalling the passages of Proclus on the bonds of Jupiter and rings of Saturn, one notes the covenant there:[185] Jupiter is the paramount god of law and order in the universe. He binds himself as well as others to obey his own laws. So the Greeks were not so far off the mode of humanity.

SEXUAL RAMIFICATIONS

Not much in the way of pragmatic life routines is exclusively male. G. P. Murdock surveyed the part played by women and men in the economic and household activities of 224 societies.[186] Only the pursuit of sea mammals and major hunting were never exclusively the task of women. But neither were they anywhere usually done by women. Nor were they ever a function shared by the two sexes. Females like the goddess Diana of the Hunt were exceptional. I attribute this not to the muscular ineptitude of women, but rather to unconscious male sexual jealousy of large beasts, and a general lesser aggressiveness in the less schizoid female, as I have explained in my accompanying volume on human nature today.

It is impossible for the human, given both the catastrophic and the physiological structure of the mind, to divorce his major concerns, to segregate them intellectually, mythically, verbally. One must impart congruity and cohesion to any important experience. He will put all experience into context, through unconscious processing by way of the catastrophized memory or through partially conscious analogizing and philosophizing. Important natural events will be related to sex, food, tools, violence, and death. Each will have, in its symbolization and in his mind, something of every other. Human sexuality is exponentially more complex that primate sexuality and reflects, with all other life-values, the circumstances of creation and the aftermath.

Truly and simply, events of the primeval period were seen to resemble hominid organs and practices. Making sense of the sky events and their

---

[185] A. de Grazia, "Ancient Knowledge of Jupiter's Bands and Saturn's Rings," II Kronos 3 (1977), 65-70.

[186] "Comparative Data on the Division of labor by Sex," 15 *Soc. Forces* (1937), 551-3.

effects, with their super-potency, called upon the new mechanisms of the mind to an ever-increasing extent, a shocking extent, until finally sexual behavior, like all other behavior, came to be a secondary derivative from imputed sky practices.

The origin of sexuality thus was attributed to the sky gods. Hence much that could relieve disaster-anxiety, the true primal fear, was forced into human sexual behavior. From that time onwards, sex was no longer the indigenous and instinctive product of the mammalian species but was the example and instruction of the gods. From the very beginning of humanity, sexual practices, like all other life and culture, were integrated and deduced from the behavior of the divine.

Therefore, those practices which in the light of humanitarian science appear to be savage or brutal were in fact instrumentally rational and functional for the new creature. Self-consciously, he and she did not care for what was natural to animals, but wanted what was possibly divine.

All manner of sexual practices and linkages of sex to other life areas came to be invented and institutionalized. In the earliest dynasties of Egypt, four-directional compass points are indicated by phalluses. Single and double-phalluses are carved as heads of batons, possibly for pointing during ceremonies and for other magical purposes; these are found in the French Upper Paleolithic sites. They are also found drawn on cave walls. Both the paleolithic Cro-magnons and the Egyptians draw a picture of heaven overarching earth: the first as a bull over a female, the second as a sister over a brother (see *Chaos and Creation,* figure 15). The drawings are so close in spirit that they may carry Cro-Magnon man down to the Old Kingdom of Egypt. A parallel occurs in Minoan Crete where Dedalus fashions a metal cow to house King Minos' wife so that she can cope with a white bull that has attracted her. These are to be interpreted not as the exaltation of sexualism as a human drive but to the divine imposition upon sex of the rule of heaven, nature, and gods.

The concept of sexual perversion dwindles when confronted by the complexity of sexuality. The Princess Palatine, wife of the brother of Louis XIV of France, in one of her frank letters describes her homosexual husband's attempts abed to fecundate her by masturbating first with holy medals of the Virgin. (The scandal of the deed and of the letter itself, which was omitted in a celebrated edition published in 1981 -- and this is scandalous, too,[187] - invites comparison, say, with the public holocausts of sinners of the same culture in church squares before the 'god of peace and forgiveness,' as described and surveyed in contemporary publications.) The

---

[187] See *L'Express,* Paris, Nov. 1981.

sacred and the profane, the obligation and the evasion, merge like wrestlers, like the symbol of the Yin and Yang. Still, public madness defines private sanity.

It is from the interpretation of divine behavior that cults of virgins and eunuchs originated and were perpetuated throughout the world. Peter Tompkins thinks that the loss of its tail by a comet identified with Venus may have originated these cults and perpetuated them practically to our day.[188] It was through the fear of the sky gods that punishment was ingrained in individual and collective behavior too, and that extremes of both impotency and furious rape came to be responses to every major and minor expression of the high energy forces.

Violence unleashed is everywhere given a sexual form and rests with the human psyche thereafter. Sex is a screen for, and release of, the primordial fear built up by catastrophic genesis and experience. No culture has escaped the process from the beginning of human time. The more obvious violent aggression associated with human sexuality is paced by sublimated sexuality. The catatonic response to disaster reflects itself in sexual frigidity and impotence, with their hundreds of individual and collective manifestations. Catastrophized obsessiveness is reflected in the frequent fixation upon the pornographic. Unlike primates, humans have developed a prolonged coitus and frequent coitus, again as types of compulsive-obsessive behavior. Human females secured a perpetual weak rut, after an instinct blockage arose against the imperative primate rut period.

Once locked into quantavolutionary theory, sexology can initiate new theories for sexual problems or deviations. The cultural relativity of sexual practices can be explained even while the universality of the catastrophe-sexuality nexus is admitted. For example, Saturnian and Bacchanalian orgies are deviant sexual as well as economic, organizational, religious, and physical (anti-hygienic) outbursts. They introduce and celebrate the end of the world in infinite series.

Once reinforced and lent new meanings by the sky gods, human sexuality entered upon the social scene vigorously. The simple dashed line at varying angles (|) is most common in cave art; it is generally adjudged to be a phallic symbol and certainly develops in that direction. The female vulva is also common (▼). The ankh (♀) symbol of (comet) planet Venus is common and may even be found in the New World as a diffused or independently invented symbol. It has been a religious symbol in Egypt and in Christian areas for millennia. It is frequently used as a genital symbol,

---

[188] *The Virgin and the Eunuch*, N. Y. Bramhall house, 1962; Zvi Rix, "Notes on the Androgynous Comet," I Rev. *Society for Interdisc. Studies* (Summer, 1977), 17-9.

bisexual or androgynous, and may be related to the Greek 'phi' (Φ) a fire sound, that was used as a sexual symbol by itself.

A variety of architectural forms have been given sexual as well as heavenly associations. The pyramid (Δ), the megalith and obelisk, the tomb in several forms, and the Greek temple are suggested as primeval sexual symbols carried into the highest civilizations. (The Temple resembles a triangular female symbol resting upon male pillars. Naked and unashamed 'lingam' and 'yoni, ' monumentally constructed, are found in India and elsewhere.

Thus may be explained the most incomprehensible of interconnections: the religio-politico-sexual. Creation events were seen to resemble actual sex organs and practices. This happens by the basic delusion that gives objective realism to signs and symbols. The superpotency of sky events and disastrous high energy forces might be controlled, it appeared, by controlling their close cousin-referents in society - sexualism. The rituals, sacrifices, elaborations, and sublimations then begin.

THE COMPULSION TO REPEAT CHAOS AND CREATION

Among the Navaho Indians, women sit on their legs, and men sit crosslegged. Why? They say that in the beginning Changing Woman and Monster Slayer sat in these positions.[189]

Manu, the Noah of India, was delegated by the gods to be the recreator of all creatures after the great flood. "In a desire for offspring he practiced worship and austerity." He "practiced severe and great self-mortification..., while he stood on one foot with his arms raised. With bent head and eyes unblinking he performed awesome austerities for 10,000 years."[190]

"We must do as the gods did in the beginning," says an ancient Hindu text. But not only the Hindus believe and act so. Every known religion does the same, whether it is the belief system of a great civilization or of an isolated small tribe.

And not only is it the religions that aim to repeat the behavior of the gods in the beginning. All social forms of activity are saturated with the emanations of this principle. In Timor, when the young rice sprouts, a specialist on agricultural myths is brought in to spend the night in the fields

---

[189] Mircea Eliade, *Myth of the Eternal Return,* Princeton: Princeton U. Press, 1954.
[190] "Manu, Ur-Napischtim, and Noah," *U. of Chicago Mag.* (Winter 1975), 10ff.

reciting the myths about the origins of cultivated rice.[191] Every activity seeks to follow its earliest principle.

The bearing and baptism of children, marriage, the rites of adolescence, and death -- sexual relations, family relations, work relations, -- governments, companies, armies, athletic teams: no activity can escape its beginnings.

Ballgames are played all over the world. They are just games, people say, and they may even say that so-and-so invented the game of baseball or whatever the ballgame is called. Not so. Every invention is in a continuity. Every game goes back to primeval religion. Every game is a game originally of the gods. The human players of athletic and parlor games are exhilarated by their unconscious replaying of divine roles in catastrophe and so are their spectators.

Thus the Olmecs of ancient central America played a ball-game and had courts built with religious carvings and paintings all around where the game was watched.[192] This was about 1500 B. C., and is attested to in recent excavations of their ruins. Their myths are clear as to what they were doing. They were imitating the games of the gods as they saw them in the sky, bloody disastrous games in which the losers, though they be gods, were killed. And so the Olmecs played their games with human skulls in the beginning, and the players who lost were killed and skulls became the balls for the next games. To shrink from these ancient practices, and take refuge rather in a supposed calm rationality of the sciences may be comforting, but is self-deceiving.

No one can escape the conduct of the gods in the beginnings. Not even the secular mind of the scientist. For even while asserting his distrust of the supernatural and legendary, the scientist uses a language, a numbering system, and forms of organization derived from the celebration of what the gods did in the beginning. Science is built upon the nature of homo schizo; it does not come from outer space. The bulk of science comes from heightened self-awareness, the wide span of human displacements, the causal connective mimicking of instinctive stimulus and response, obsessive attention, suspiciousness, associations retrieved by naming, and imitating with arithmetic addition the sequential processing in the consciousness' (dominant ego's) control of attention. Scientists constitute a corps of disciplined self-controllers engaged in these schizoid practices.

---

[191] Eliade, *op. cit.*

[192] Carmen Cook de Leonard, "A New Astronomical Interpretation of the four Ballcourt Panels at Tajin, Mexico," in Aveni, ed., *op. cit.*, 263-83.

Furthermore, like a snail moves with its shell, the scientist carries his shell of culture as he goes about his work.

An Einstein will trust that "nature does not play dice," a scientific fiction that perhaps is not as reality-based as the cosmic fiction of early man, who went back to the beginnings, when the gods were playing ball. In India the game of dice may have begun, say Santillana and Dechend, with the gods, "who go around like... casts of dice." [193] Indeed nature plays dice. The Hindus also played a game called 'planetary battles. ' 'Nature, ' of which Einstein speaks, is an idealization of Zeus, who maintained law and order, despite his rapscallion son, the planet-god Hermes (Mercury) who was always traveling about and bringing luck to dice-throwers.

What causes this compulsion to connect all activity to its origins in the primordial conduct of the gods? A compulsion to repeat an event, say the psychiatrists, of whom Freud may have been the most eloquent and original, is caused by the traumatic nature of the event to those who experience it. What, then, was so shocking about the events of the beginning? And how can we be traumatized today by events so ancient that they slip off the pages of recorded history?

Now again, unanimously, the fossil voices of the dimmest past speak; the events were the awful behavior of the gods when they created mankind. They tore apart the elements of nature to fashion this new creature. They drenched the world; they fired it; they tore up the earth; and they stormed the atmosphere. The human creature was made from the elements in a time of great stress. He was born to great fear and abject servility to his makers. He was born with a compulsion to repeat his birth throes. The birth of every infant, as in the theory of Otto Rank, is the primal trauma of the person; the birth of mankind is the primeval trauma of humanity.

SUBLIMATION

I explain in an accompanying volume my doubt that the word 'sublimation' is scientifically useful. It refers always to a displacement; hence what is said about the one is to be said about the other. Therefore, while I retain the word in these passages, the word 'displacement' can be read into them equally well.

The gestalt of creation inaugurated for the new person a kind of incessant civil and foreign conflict, one in which his resources were over-extended internally and externally. He could not keep himself in order

---

[193] *Op. cit.*

without ordering the world outside, a generally impossible task but one to which he was now biologically committed. But, as it happened, his most effective ally was his external enemy, society and culture, including gods. These are all non-existent delusions, and hallucinations when they 'command' one, but nevertheless interpose true ordering principles of conduct into the behavior of the inveterate warrior. In effect they tell him that he must 'sublimate,' and teach him how to do so.

Sublimated behavior is commonly understood as an unconscious substitute in socially accepted form for impulsive behavior that would be condemned. The substitution appeases the unconscious while it performs its overt function. Primevally, sublimation begins in the overt function. Primevally, sublimation begins in the distraught circumstances of self-awareness, as amnesia and recollective memory begin.

When the anxious species is born and asks of itself an impossible measure of control, it begins by reacting against the memory of its harsh experience, which is soon submerged ('forgotten'). The memory is still active and sets up a ghost pain against the recurrence of the experience. Consequently, the memory lends itself in translated and disguised form to unrecognized expressions. It is carried back to the surface of the thought and activity, where it is enacted and reenacted in disguise. It is used in a solemn rite, or played with as a toy, a game, or a comedy. Thus, as was described, ball games became sacred, dramatic events, which reproduced the battles among the celestial hosts, with skulls as balls, and beheading as the price of defeat. As in heaven, so on earth. Control is established; the memory functions in a displaced setting, exuding the energies of the response that it demands but cannot perform frankly.

Only a theory that human nature is schizotypical can explain the vast and ramified character of sublimation. Students of language, myth and art, in their diligent search for principles, have discovered that myriad delusive and distorting guises can surround any event. Yet sublimation occurs not only in linguistic and artistic life-areas, but also in all areas of technology.

The original traumas and mental distortions of humans required all things in the objective world to be processed through the schizoid world and there given some of their meaning and forms. Beautiful church liturgies, children's fairy tales, methods of combat, designs of tools, and systems of philosophy abound in examples.

Freud once wrote in Totem and Taboo that "the neuroses on the one hand display striking and far-reaching resemblances with the great social productions of art, religion and philosophy, but on the other hand they have the appearances of being the caricatures of them. One might venture the statement that hysteria is a caricature of an artistic creation, the obsessional neurosis a caricature of a philosophic system."

In all three cases, the reverse is more accurate. The nomenclature is irrelevant. For instance, hysteria is regarded here by Freud as a poor artistic creation, whereas actually the art is a sublimation of hysteria. All three highly regarded sublimations are founded upon the primordial madness that says in its first gestalt: fear, remember, control; and, when external controls move against one, sublimate!

In the eight century B. C., Hesiod wrote a combined philosophy, theology, and poem, containing what may be elements of a correct cosmogony. Gods and muses and humans transact in a highly metaphorical and figurative drama. Between the reality and his mythology lay an enormous collective pain, felt by a people who had not better way of confronting the terrors of existence and the traumas of their history. It is so, too, when a schizophrenic patient gives a fully pseudo-mythical account of an event that contains within it an accurate report that he is too pained to tell about 'as it really happened.' There is a sad irony here that the more he succeeds to sublimate, the worse the diagnosis of his illness. But is he not a 'pathological liar'? He is that, too, except that his lies are embellished in a fashion close to what society will accept as myth or as a work of art. When Gustav Mahler composed the "Song of the Earth," he was a neurotic who was contemplating suicide, but meanwhile he was also communicating to his audience, the society, a message in a modified 'modernized' language that they would grasp on the brink of their own madness. His song, his neurosis, his suicidal intents - these were all himself trying to cope with his depersonalization; but the audience would say, 'somewhat mad, but sublime.'

Shakespeare has joined together the transacting elements in a few lines of *A Midsummer Night's Dream:*

> Lovers and madmen have such seething brains,
> Such shaping fantasies, that apprehend
> More than cool reason ever comprehends.
> The lunatic, the lover, and the poet,
> Are of imagination all compact:
> One sees more devils than vast hell can hold,
> That is, the madman; the lover, all as frantic,
> Sees Helen's beauty in a brow of Egypt:
> The poet's eye, in a fine frenzy rolling,
> Doth glance from heaven to earth, from earth to heaven;
> And, as imagination bodies forth
> The forms of things unknown, the poet's pen

Turns them to shapes, and gives to airy nothing
A local habitation and a name...

What we would conclude here is that sublimation is but whatever is socially acceptable, now or later, in homo schizo's methods of handling his displacements. Everyone, not artist and scientist and humanitarian alone, sublimates at work and play, in love and indignation, in common speech, yes, in dreams asleep and awake. One must sublimate, and always has sublimated since the earliest generations when a *modus vivendi* had to be established among the schizo clan.

CANNIBALISM

A common textbook example of sublimation was provided us by William James, who suggested that sporting contests functioned as substitutes for warfare. A weaker surrogate is afforded the spectators, and perhaps even weaker is that tendered to the masses who watch the sports on television. The problem of warfare is much more complex of course.[194] The infantryman usually hates war. Why should he need a substitute? Whereupon we begin a painstaking unraveling of the web of war, attacking the knots of displacement; projection; paranoia; aggression; tradition; authority; habitude; greed; loot; rapine; prestige; exhilaration; gambling; remote and indifferent pilots, missiles, and generals; racism; economic competition; self-destructiveness; the Armageddon-complex; and so on until all the knots are untied and not much is outside of war except a few rules of the Geneva Convention which beg us not to kill prisoners, and certainly not to eat them, as now and then has been the case. Achilles wanted to eat his slain enemy Hector and Hector's mother, well, she wanted to chew Achilles' liver.[195]

Cannibalism has also had to be sublimated, rendered by frontal cultural attack into a taboo in most cultures, a nauseating abomination, but then also built into the most symbolic and elaborate rites, as in the Eucharist of the Christian Catholic religion.

The suppression of cannibalism must be one of the most successful and important sublimations that mankind has ever achieved. Cannibalism is not unnatural to humankind or else it would not seem so repulsive and dreadful.

---

[194] *Cf.* Quincy Wright, *The Study of War*, Chicago: U. Chicago Press, 1965.

[195] Eli Sagan, *The Lust to Annihilate: A Psychoanalytic Study of Violence in Ancient Greek Culture*, N. Y.: Psychohistory Press, 1980.

Further, it would not be so commonly discoverable in sublimated form among religions in the world. The major question is, in fact, not whether, but how we came to be numbered among the rare species who have eaten their own kind.

The theory of homo schizo here offers three reasons. The first is that a creature that can fear and hate itself, and can transfer this ambivalence to others and gods, can sternly rationalize the eating of others, which symbolically includes itself and the divine. Second, the practice, which has psychic and religious justification, has had upon many occasions a pragmatic or calculated effect; people who would otherwise starve upon the occasion of near extinction from natural disasters, including famine, flood, and radionic plagues, would eat whatever came to hand. Third, relations for some time with others of one's band and tribe would include a stratification between homo sapiens and hominids. A logic of divine ritual sacrifice, made urgent by protein starvation, could be confirmed by primeval considerations of eugenics, population control, animal husbandry, and invidious racism. Cannibalism was restrained and sublimated very early because it was self-threatening; one was ingesting other egos of an uncontrolled kind except under rare stable ego conditions. As with psychogenic mushrooms, you have to be a bit crazy to eat them and you become distinctly crazy afterwards.

Anthropologists have long suspected earliest humanoids of cannibalism. From Ethiopia (Valley of the Awash River), the Bodo hominid skull has, upon reexamination, been adjudged a victim of ritual defacement and scalping at least, with a probability that it had been treated anthropophagously. [196] These operations which use tools, are human, whether perpetrated by Bodo man or by related hominidal or human types.

Nowhere to our knowledge was cannibalism more widely practiced than in the Aztec empire prior to the Spanish conquest. The Aztecs placed "war and its corollary, sacrifice, at the very center of their universe... the one and eternal order."[197] Tens of thousands of prisoners were taken, nourished, sacrificed, and eaten every year.

Gert Heilbrunn calls cannibalism "The Basic Fear," and writes that the infant is born cannibalistic and projects its impulses upon the environment as his persecutor. "Phylogenetic and human ancestral reflections in

[196] Based upon paper and discussions at the American Anthropological Association Convention, Atlanta, Georgia, August 15, 1982, led by Donald Johanson and Timothy White, with supporting statements by Louis Binford, Glenn Conroy and Clifford Jolly.

[197] Burr C. Brundage, *The Fifth Sun: Aztec Gods, Aztec World,* Austin: U. of Texas press, 1979, 217.

conjunction with psychoanalytic data point to the ever-existing threat of
passive cannibalistic incorporation as the basic danger" felt by the new
organism.[198] He finds cannibalism widely spread among historical human
groups and sublimated very often in modern groups.

Let us now turn to anthropophagism in its most sublimated form. We
begin with the warning of St. Paul (Epistle to the Corinthians, XI, 29): The
Christian who enters upon communion without comprehension "eats and
drinks his own damnation." For the Christian is partaking of Christ. The
text of the Gospel of John recites the startling words of Jesus, which
shocked even his disciples, and this text is repeated by the priest at the
moment of Consecration in the Mass:

> Most truly I say to you, unless you eat the flesh of the Son of man and
> drink his blood, you have no life in yourselves. He that feeds on my flesh and
> drinks my blood has everlasting life, and I shall resurrect him at the last day; for
> my flesh is true food, and my blood is true drink. He that feeds on my flesh and
> drinks my blood remains in union with me, and I in union with him.(6: 53-6)

Henri Fesquet explains what occurs in the Eucharist:

> The communion, is it cannibalism? To judge by its intent it is undeniably
> so. It proceeds through a murder, a sacrifice, through manducation, and
> through the classic symbolism: uniting a person with another whom one
> loves, and appropriating his qualities. To eat God is to make oneself
> divine. But the sacrament is more than cannibalism. It surpasses and
> sublimates it. It is disconnected materially from the cruelty of the killing,
> granted that, without Golgotha [the mount of crucifixion], there could be
> no Eucharist, which directs the separation of the flesh and the blood.
> Besides, the raw material of the Eucharist, the bread and wine - two
> products of the earth - gives it a cosmic dimension, actually a pantheistic
> one. The vegetable kingdom, among other things, precedes the animal
> kingdom and, in a sense, engenders it; by means of the Eucharist, the cycle
> of creation begins once more. That the presence of Christ is total (" real"
> in the bread and the wine as Catholic theology maintains) gives to the
> incarnation an exquisite prolongation and deprives the embodiment, the
> cannibal effect, of all its cruel character. Here, the violence of love is made
> silent, decent. The consecrated hosts and cups of wine, these there will
> always be, everywhere and for everyone. It is the superior gesture of
> tenderness. The mean which Jesus conveyed to his friends achieves a
> universal character. It is the virtue of Christianity, which traversed the
> grounds of the religion that came before it, to have adopted the better of

---

[198] 3 *Jour. Amer. Psychoanal. Assn.* (1955), 44-450, 464.

them, to have purified their rites and broken down the barriers of races and nations.[199]

The Eucharist might be compared with an entry in the Notebooks of Leonardo da Vinci where he wanders from his point, which is to do honor to virtuous persons. They

> deserve statues from us, images and honors; but remember that their images are not to be eaten by you, as is done in some parts of India, where, when the images have according to them performed some miracle, the priests cut them in pieces, being of wood, and give them to all the people of the country, not without payment; and each one grates his portion very fine, and puts it upon the first food he eats; and thus believes that by faith he has eaten his saint who then preserves him from all perils.[200]

Collective violence need not proceed across territorial boundaries. The killing and eating of members of one's own group, particularly of those somewhat different, physiognomically and mentally, and without foresight to arm and attack first, could occupy a normal place in the career of the homo schizo band.

We recall that the deceased and digested are often relatives, at the least of very similar kind, although, for that matter, there would be a general animism operating by displacement and projection to make many mammals one's relatives and even totems. The general animism would only serve to make the killing and eating of one's own kind less remarkable.

At what point would the practice cease and guilt be felt? For cannibalism, soon; for war, never. As evident hominids diminished in number, homo schizo would find himself battling with and dealing with his speaking and aggressive kind almost entirely. Identifications would become closer and closer until one was eating "oneself." This should stop most cannibalism. Further the years of easy human hunting would be over; other animals were easier to kill; improved weaponry played a part in this switch of practices.

Now then the cannibal victim can be identified with oneself (seeking esteem) and one's gods (requiring sacrifice). The outcome would appear to be sacred cannibalism, reserved more and more for ceremonial occasions and anniversaries. Thereafter any promiscuity in eating human flesh would

---

[199] Henry Fesquet, "Anthropophagie, sacrifices humains et immortalité," *Le Monde,* June 21-2 1981, l, 17.

[200]Quoted in K. R. Eissler, *Leonardo da Vinci,* London: Hogarth, 1962, 262.

become tantamount to a crime against the gods and spirits; a taboo would be born.

Prisoners were first sacrificed and eaten, then simply sacrificed. Ultimately animals were sacrificed and eaten; this practice would be less thrilling but more reliable, since, for other reasons, prisoners of the right kind were not always available. Like infant sacrifice, prisoner sacrifice is more disturbing when other means of controlling the gods are less threatening to poly-ego stability (I must stress that this poly-ego stability is not an absolute, objective state; rather it is a cultural balance uniquely fashioned with the individuating traits of the person.)

VIOLENCE AND WAR

Meanwhile collective violence continued unabated, enhanced indeed by social growth. Primates do not wage war. Female primates do not even kill game. Baboons fight, and one band will live apart from another, trespass upon another's territory, and engage in a melee when challenged; instances of a baboon being severely injured or killed in such warfare are rare. Strange individuals are usually chased away by a band's 'citizenry, ' but on occasion adopted.

They do not foresee war, plan war, reenact or celebrate the anniversaries of war, or train for war. The hominids behaved like primates. Warfare is peculiarly human and naturally emerges from the schizoid traits of self-awareness, memory, group-shared symbols, projection, and the limitless search for the impossible goal of control over the self, gods, others and the environment, staked by endless fear. Aware of oneself, and fearful of it, the person recalls the creators in subservience, propitiation and terror. As the creators do, so does he. The wars of the gods have always been in his mind as models of behavior. Even Christians carry along a war of god and the devil, descendents of Horus and Seth in Egypt, Jupiter and Lucifer (Phaeton) in Greece and Rome, Tezcatlipoca representing both in Aztec Mexico, according to Brundage.[201]

If culture appeared promptly upon humanization, then warfare and other forms of collective violence may have roots in the same process. I say 'may' rather than 'must' because warfare and other human practices might be considered most important as effects, yet maintain no essential connection with the dynamics of human nature. If such were to be the case,

---

[201] *Op. cit.*, chap 4.

we should rejoice, inasmuch as collective violence might then be more readily extirpated from culture.

It is pointless as well as impossible to survey here the voluminous literature on human conflict from several major scientific fields.[202] Suffice to say that no culture, anywhere, is surprised at collective violence, whether internal or external, and, to that degree, a natural quality is indicated for it. However, nowhere does the practice of collective violence bring on, except in individual cases, the physical revulsion that cannibalism often excites when it is experienced or reported. Taboos against taking up arms, for 'just' cause of course, are rare.

Ethologists, naming K. Lorenz, Arbrey, and Björn Kürten as instances, often claim that man was originally a cannibal warrior. Iränäus Eibl-Elbesfeldt finds warfare in prehistoric societies and in hunting and gathering cultures today.[203] Kürten's theory, that man was originally two or more kinds of primate of generalized brain and instincts who struggled with each other, gives us a lead to pursue. An early human band, composed of homo schizo or dominated by the type, would be in continuous contact with unaffected hominid bands. The culture gap between the two species would be wider than their appearances might suggest. Even though cultural assimilation had to recommend itself to homo schizo, and he tried in the earliest times to accommodate hominids in his 'table of organization, ' there would occur internal rebelliousness and flight. The neighboring hominid bands would have no means of understanding nor wish to learn that they might become docile enough to appease homo schizo, and find a place in his expanding territory. Thereupon, they would become targets of aggression by homo schizo, who would kill some, break up the band, and acquire the more docile as slaves who would join his 'breeding farm. '

Successful violence encourages more violence. So the practice of war would be common and energetic. Relative to other sources of labor, sex, breeding, and control, bands composed entirely of hominids or almost so would be most lucrative. And as we said above, they were an excellent source of food. Peking man, homo erectus, along with other types of man, have presented some fossil evidence of cannibalism. The food would be both hunted and farmed.

If the human felt at ease with himself, whether or not he controlled the world, it is doubtful that he would so persistently engage in the most risky of enterprises -- collective violence. Enough has been said in this work and elsewhere to stress that the problem of 'feeling at ease with oneself' is no

---

[202] Q. Wright, *op. cit.* W. D. Hamilton, in Fox, *op. cit.*, 133ff.

[203] *The Biology of Peace and War*, 1975, trans. N. Y.: Viking, 1979.

matter of a decent meal and a good night's sleep, but it is the greatest and most persistent human predicament. Contributing to its recalcitrance to therapy is its embodiment in the central nervous system, where to suffer and inflict suffering is tolerable and even appeasing and the urge to control extends beyond sight, beyond the grave, into the skies.

C. G. Jung would make of this the eternal celebration of a destructive archetype, developing out of the eternally split condition of the soul. The archetype is "that which is believed always, everywhere, and by everybody," and if it is not recognized consciously, then it appears from behind in its "wrathful" form, as the dark "son of chaos," the evil-doer, as Anti-christ instead of Savior - a fact which is all too clearly demonstrated by contemporary history. [204] Moreover, the natural storms amidst which hominid was mutated into man and which occurred throughout his earlier history added to his fright and stressed his already biologically catastrophized nature. In every disaster some people run about proclaiming the work of the "Evil-doer" - "The Evil One," one Alaskan 1965 earthquake survivor called him. All institutions and cultural practices are permeated by natural catastrophes. Their effects upon mankind would be fully apparent in history and psychology were it not that mankind already displayed and carries similar effects in his nature - 'white on white, ' disaster upon disaster.

---

[204] "A Psychological Approach to the Trinity," in *Psychology and Religion: West and East, II Collected Works,* New York: Pantheon, 1968, 117.

CHAPTER VII

PSYCHOPATHOLOGY OF HISTORY

The death-scream of Lady Macbeth is heard off-stage and Macbeth, told of her end, generalizes the human tragedy:

> Tomorrow, and tomorrow, and tomorrow
> Creeps in this petty pace from day to day,
> To the last syllable of recorded time,
> And all our yesterdays have lighted fools
> The way to dusty death.
> Out, out, brief candle!
> Life's but a walking shadow, a poor player
> That struts and frets his hour upon the stage
> And then is heard no more.
> It is a tale
> Told by an idiot, full of sound and fury,
> Signifying nothing.
> (Act V, *scene 5*)

Whereupon he sallies forth to battle; death is the therapy: *il se fait tuer* or, as Americans express it, 'he gets himself killed, ' Life" is a tale told by an idiot," the many idiots who live and then those who tell of it, and such is history. These famous lines of Shakespeare seem to be in context here.

Starting with its creation, mankind moved through time on a spiral path around its schizoid core. On numerous occasions catastrophes changed the arc of the spiral, sending humanity closer to the core in mentation and behavior. Whenever the natural environment seemed to settle down, it appeared that he might invent ways of reaching beyond his limitations, and his historical spiral moved away from the core. But simultaneously, as if magnetized by the core, he would be pulled inwards to it. Thus it has happened that the record of some five thousand years of proto-history and history has found mankind reenacting time and time again, without the urgency of catastrophes, his primordial behavior. His spiral path has had an inertia such that he could neither escape his core self, nor the fossil thrusts of the disastrous times that he had suffered.

When a social change occurs, when the earth trembles, when a comet flies by, his mind is unduly disturbed, that is, agitated beyond his normal schizoid behavior into activity reminiscent of the similar but much greater catastrophes of his earlier days on Earth. It is not necessary, either, to search only in the wreckage of recent disasters of relatively large social scope for an outburst of symptoms of schizophrenia. Even in mild episodes, whether collective or individual, the symptoms exaggerate, become full-blown. The recurrence of such symptoms, great or minor, in modern times, are not to be thought of as evidence of the weakness of the quantavolutionary model of homo schizo, but, on the contrary, are indications of its strength.

Let us examine a few types of historical behavior to clarify the 'psychopathology' of history as the story of homo schizo. When we do so, we can agree that Arthur Koestler, man of much political experience as well as a profound human analyst, was hitting close to the truth as he was writing: "When one contemplates the streak of madness running through human history, it appears highly probable that homo sapiens is a biological freak, the result of some remarkable mistake in the evolutionary process."[1]

A SICK JOURNEY

Generally, history-telling is a guided use of symbols to integrate disordered minds by reiterating obsessions about how the idols of the tellers

---

[1] *The Ghost in the Machine*, N. Y.: Macmillan, 1968, 266.

have controlled the world on their behalf. So history-telling can be a form of dance, chant, prophecy, folklore, legend, prayer, catechism, rite, epic poem, parable, fiction, literature, fantasy, sacred relics, drawings, sculptures, physical constructions, or historiography in its narrow sense. In the total accounts of humanity, historiography, which looms large in the minds of such as read this book, is a small part of history, with only a fractional impact upon the total effects of history-telling.

It is a history-telling when people engage in sacred drama or dancing, as all drama and dancing were during most of the world's history. Nor does comedy or jazz dancing or even computer music escape its sacred roots, although these are sublimations of sublimations beyond facile recognition. Sacred dramas have occupied more human time in history than the whole of all secular theatrical activity since its beginning. It is common to find 'madness' in Shakespeare and Samuel Becket, but a sense of direct connection with primeval origins does not come readily. Let us use the materials so auspiciously gathered by Perry to stress the interconnections of dance forms, schizophrenia, and the origins of mankind.

Persons going through psychotic episodes frequently say that "they are taking part in some dramatic performance that has been already written and prepared beforehand... nothing about them feels arbitrary or 'made-up, ' but rather they seem to follow well-established configurations."[2] They are linked actors in ritual dramas found around the world - Egypt, India, China, early Central and South America, the Norse area, Ireland, Iran and other Indo-European regions.

The inner journey of the psychotic topically repeats the following form, in the words of Perry:

1. Establishing a world center as the locus.
2. Undergoing death.
3. Return to the beginnings of time and creation.
4. Cosmic conflict as a clash of opposites.
5. Threat of the reversal of opposites.
6. Apotheosis as king or messianic hero.
7. Sacred marriage as a union of opposites.
8. New birth as a reconciliation of opposites.
9. New society of the prophetic vision.
10. Quadrated world forms.

---

[2] Perry, *Roots of Renewal in Myth and Madness,* 80.

Carrying this framework into the ritual drama of Ancient Egypt, we encounter the replication of raising of the center of the city as did the creator god. Next there occurs the drama of the murder of the god Osiris, and of ensuing chaos, until his son Horus assumes world power, and is embodied in the Egyptian monarch. The king goes back to the beginning of creation and is baptized, purified, and prepared for the future. The devil god Seth, represented by an animal, seeks to usurp the monarch; mock battles are fought with warrior actors. A suspenseful period is said to follow, a chaos, while the issue of ruling the universe is unsettled. The king is crowned ruler of the world, successor to Horus. Various festivals are held; effigies of the gods cohabit. Osiris is reborn through his mother, the Sky. A new harmonious order of the world is proclaimed, reestablishing the primal order and justice. Throughout, the world is represented by a four-cornered, four-pillared structure, the four cardinal directions. The king takes possession of them all. Although a jumble of celebrations and their related dramas develop in Egypt, as elsewhere, and do so in the varying forms that personal psychoses take, the general paths of the rituals remain clear and it is likewise strikingly evident that the great society is celebrating a thoroughly schizoid cycle, year after year, endlessly.

The mélange of ritual dramas adds up to history, as it is known and relived by the elite and masses of all times and places. History as it is taught in the schools, schizoid though it may be, is but a pallid imitation of this more fundamental and archetypical history-telling. The Exodus story, true in most respects - as indeed the drama contains essential historicity for all peoples - is recited and replayed in Judaic ritual celebrations. The Roman Catholic mass, basing itself upon the life of the Christ, is also a ritual drama. Joseph Campbell composes a worldwide plot for the tales of heroes, which belongs in the category of ritual dramas.[3]

What happens to the educated and scientific people, the millions of unbelievers, 'back-sliders, ' the communist throngs of half the world that denies the ritual drama in its traditional forms - do they successfully cast off the schizotypical behavior implicated in the ceremonial dramas? More likely, they find substitute outlets. They pursue speculation on the origins of the universe (the catastrophic 'Big Bang'), on the evolution of life over billions of years, on the climactic but prolonged rise of mankind, on blind but progressive nature and on its control by reason. In a recent television film, Cosmos, viewed by millions and loudly touted by the intelligentsia, Carl Sagan, an astronomer, chants a prolonged liturgy on the evolution of life

---

[3] *The Hero with a Thousand Faces*, Princeton, N. J.: Princeton Univ. Press, 1949, 88.

forms from molecule to man, with the help of canonical background music and sleight-of-hand cartoons.

Bits and pieces of the ritual drama (which was not a one-act performance anyhow) are parceled out to holidays, parades, 'Hollywood westerns,' speeches, diet-fads, paranoid political causes, mass spectator sports and so on. Admittedly the sundered great god Osiris is rarely discovered and put together again. Living a new history, as contrasted with reenacting faithfully history, is difficult. But for those who cannot stand the secularized way of life, there is then mental therapy: 'Maybe you should see a psychiatrist, ' or "There's an article you should read in yesterday's newspaper".

HISTORISM

History-telling today is typically viewed as the events of the past, incompletely or completely related depending upon the number of volumes given over to it, sorted by periods like the Renaissance, names like Julius Caesar, topics like architecture, events like the Battles of Verdun, or demonstrations of principles like Marxism. Behind all of this is the historian: "Alexander is Great," said his faithful biographer, "because I wrote about him."

What do historians write? If what history tells us is true, then homo schizo is the hero of all times and places. If what history says is false, then history is the workings of the minds of homo schizo on past events. Should someone protest that history is both true and false - and indeed it is - then homo schizo must be both subject and author, and then in a way all history is the autobiography of this species.

Historism is a production of histories about history. In its various guises, it is evoked in order to train the people of a culture how to avoid and handle anxieties. It is typically addressed to some part of society in preference to the rest because there are usually several identities striving for recognition of themselves and no history can or wants to work for all of them. Ordinarily it is the rulers who have most need of history and can command it most readily and can support it. It appears no less than right that the 'heads' command the 'heads.' It is clear that historism is a branch of culture, a culture complex, and little more. The two have the same motive, to help homo schizo behave in a controlled manner.

The major focus of historism is in a fundamental sense upon itself, that is, upon the power needs that brought it into being. What are the main problems here? Historism must show how first came chaos, then the creation of the forces that work for the benefit of its sponsors and clients. So historism must deal with the creation, with the gods who are its gods, who have chosen its clients. Then it moves to the rulers, its rulers and how they worked for its clients until, usually, the rulers went bad. It unites fate and destiny with the clients, but if the gods are not active enough, and if fate and destiny are not adequate, then it adjoins some lawful principle, like evolution by natural selection, like progress, like the triumph of the working class, or like the ideal of the nation-state.

Creation, gods, and rulers or principles - these are the major subjects of history, the events of history, even written history, much more oral history. Now then historism fattens itself into great tomes, as in epics, encyclopedias and monographs, to concentrate upon the settings or conditions of different times to make certain that all clientele will have a locale and moment with which more easily to identify.

Then historism concentrates upon conflicts, and as in peek-a-boo with a baby, which Otto Rank sees as a basic play for relieving the fear of a separation from the guardians, the conflicts go to show how time after time the ego's stability is threatened by accident or malefactors, only to be restored by benign and usually anthropomorphised agents. In the end the topics of history, the main topics, not the endless sublimations of topics, serve to concentrate attention and relief where it most matters, at the most threatened points: why we are here in the first place, who we are, what has been done to keep us reassured - even by the most devious means - and how we may expect to preserve our being into the future.

Historism, then, in all of its forms, is therapy on a grand scale for homo schizo. It operates like a giant brain. It helps a selected main ego to dominate the other egos. It occupies itself with the displacements that are current and molds them into more meaningful buttresses of the self, so at one time it concerns itself with gods and then at another time with heroes and rulers, then with food supplies, with money, with ships, whatever the focus of the attention of its clients.

It aids the memory to forget and recall. It says to the Jews of 3400 years ago, 'Remember your bondage in Egypt, ' and to the Jews of 1981 'Remember the Nazi holocausts.' And the Romans said 'Remember Carthage' and the Americans said 'Remember Pearl Harbor.' One may wonder that these are disasters. But the disasters are still the route to victories; the end has not come. Meanwhile they give goals, *ergo* identity, to the clients of historism. The Germans of 1980 do not proclaim 'Remember World War II!' with much enthusiasm; they are seeking a new dominant ego.

If the disaster is final, it is suppressed. The American Indians and Blacks, for a long time, wanted no history and knew hardly any. It was ego-destroying. If, by some concatenation of events, it is revived, then its more tolerable parts are recalled. The unbelievable catastrophe that Elohim brought upon the world in the great flood is so honeyed for the surviving clientele, for the Noahs, that it can be read by tender-hearted little children without qualms. Historism supplies the proper amount of amnesia.

In addition it distorts or even denies the events. Even sincere efforts of German policy since World War II have not prevented massive amnesia of the death camps, and, for more complex reasons, German democratic leaders have had to tolerate deliberate efforts to show that the Nazi holocausts were unknown to most Germans and also greatly exaggerated.

All the perfumes of Arabia could not sweeten of murder the hands of Lady Macbeth, the poet wrote, and so guilt goes underground and historism is often nothing but sublimations of crimes perceived and committed. It becomes part of the devil within one, the uncontrollable alter ego and helps homo schizo to remain himself, eternally divided, and therefore adds to the continual flow of anxiety, maintaining the level that guarantees the genetic predisposition to remain an unstable self.

If it is true, as psychoanalysts say, that one has to be psycho-analyzed before he can practice psychiatric therapy, then it must be equally true that all historians should be psychoanalyzed. But if this were so, then history would be a dull compendium of who knows what; or it might not even exist, for of what use is remembering if it does not tender oneself a psychic strength? Conceivable, but impossible: if it were done, all poetry and history and literature and music would be lost; it would be irrelevant, dead, unpracticed. Or perhaps the psychoanalyzed historians will be told by their therapists what their age-old mission is: not truth, but therapy.

But homo schizo is quite incapable of this, although he toys with the idea as we play with it here. Historism gives him control, relief from fear, a more manageable ego, comfortable obsessions, paranoia and aversiveness, cognitive disorders quite believable, grounds for ambivalence, negativism; and above all it affords him sublimations. Verily, history is too important to entrust to truth.

SCHIZOID EPISODES IN ABUNDANCE

The human never acts according to a single factor in his complex, but in terms of the complex itself. Whence history as a whole can be viewed as a prolonged struggle against anxiety, as Norman Brown, for example, asserts.[4] Action according to a single mode, e. g., 'obsessive compulsion, ' without involvement in identity questions, displacements, fear-level, or

---

[4] *Life Against Death*, N. Y.: Vintage, 1959.

sublimation, does not occur. Rather, the history of a set of actions, of a character, or of an institution involves all modes in different proportions and with intricately woven and sometimes imperceptible patterns.

Technical and scientific histories, say defenders of objective historism, are exceptions to the flow of schizoid control processes through accounts of the past. Firstly, as has become accepted by historians of science, in principle if not in practice, such specialized history has the full range of homo schizo behaviors in its substance. The writing itself, far removed from a chant about the first days of creation, is a subterfuge, proclaimed so openly and therefore deemed innocent. The sublimation of factual technical narrative, even in its suspiciously professed narrowness, is intended to follow an obsessive rhythm, letting all the faculties of homo schizo sleep and dream while the brain beats to a narrow band of 'truth. '

Chess is a highly intellectual game. Computers can play it close to the master's level. Below is a story that may not be in the history books of chess, it being counter to be rationale of the game.

> An extraordinary everyday-life example of a paranoid reaction illustrating shame-humiliation mechanisms took place at the Spassky-Fisher chess-match of 1972 held in Iceland for the world championship. By the 17th game, Spassky, the Soviet world champion, was facing the loss of the match by three points, 9 1/ 2 to 6 1/ 2, with 12 points needed to win. In this symbolic warfare between the Union of Soviet Socialist Republics and the United States, a humiliating defeat was impending for the Soviets, who had held the world championship for the previous 24 years and for 41 of the last 45 years. Soviet chess was about to lose esteem in the eyes of millions.

> Trying to account for Spassky's "unusual slackening of concentration and display of impulsiveness," the Soviets issued a public statement claiming that non-chess means of influence (electronic devices and chemical substances) might be involved. They requested an expert examination of the playing halls and its contents.

> A 24-hours guard was placed around the hall. The chairs of the players were examined for poisons and x-rayed. (These were richly upholstered Eames chairs sent from New York.) Icelandic scientists dismantled the lighting canopy over the stage, but found only two dead flies. (Strangely, no one suggested further examination of these well-known bugs.) Nothing was found.

> At the end of the match, which Fisher won 12 1/ 2 to 8 1/ 2, Spassky, who during play sometimes peered suspiciously up at the lighting, said, "I

still feel there was something in the hall that affected me... I am really convinced there was some curious thing in it."[5]

Whence, for that matter, come the highly elaborated practices of medical therapy?

In the second millennium B. C., the Chinese word for "medicine" still was composed of two parts, "cure" and "divination." W. Tseng reports that the concepts of Yin-Yang opposites, of the five elements, and of the microcosmos-macrocosmos bond dominated the *corpus medicus*.[6]

Care and feeding of the young were perhaps the earliest therapies. Care of the self has been noted among some mammals. The licking of wounds is common; they may even be bathed. Care of other adults of the group is found in warning signals, grooming, and food-sharing. This, if it were not reflexive or conditioned, but voluntary, would require the multiple identification process of homo schizo. Worship of gods implies care and attention to the projected demands and needs of the controllers upon whom one's sense of self-control depends.

The first medical therapies, it may be conjectured, were reiterative rites and celebrations, such as dramatization of big dreams, orgiastic feasting, cannibalism, self-mutilation, mud baths if it was believed that we were fashioned from the primordial ooze; blind staring of catatonia; emitting sounds evocative of pandemonia; and hypnosis. Recapitulation of collective trauma, of natural disasters and defeats, was foisted upon the group as a mode of therapeutic control. All of these are found today in highly altered forms as mental and physical healing.

Then might proceed the infinitely varied and slightly less 'mad' corpus of homeopathic medicine. Displacements occur by gestalts far removed spatially from resembling gestalts in the brain. The essential methodology is still reiterative, but one large step removed to the imitative by means of extended analogies - exploring the overlapping discs of the neuron nets for discovery of what new connections make one feel better. In the homeopathic mood a therapist might readily move into the finest sublimations, recognizable as to their origins only with difficulty.

Eating the lotus flower is far removed imaginatively and practically from sacred castration as a way of controlling the god of a comet or a planet like Venus, which is associated sometimes with the lotus and with castration and

---

[5] K. M. Colby, "Clinical Implications of a Simulation Model of Paranoid Processes," 33 *Arch. Gen. Psychiatry* (July, 1976), 854-7, 855-6.

[6] Wen-Shing Tseng, "The Development of Psychiatric Concepts in Traditional Chinese Medicine," 29 *Arch. Gen. Psychiatry* (October, 1973), 569.

clitoridectomy. The effects on a wound or on an aberrant mind are achieved, whatever they may be, at the same time as the cooperation (control) of the god is achieved.

Ultimately the development of a set of plants, usable in a variety of complaints, is recognized, accepted and even experimentally enlarged. All this now occurs beneath the sublimatory umbrella of suppressed, 'forgotten, ' religious approval; thus, the god of Venus may rarely be evoked or cited. A corpus of medical therapy exists and can even grow pragmatically by means of the observation of qualities, doses, and effects. Homo schizo has no objection in principle to actual cure, so long as the cures are by-products of or do not interfere with self-control.

For lack of space, the interminable parallels (really homologs) between schizophrenia and archaic human behavior cannot be drawn out. The invitation is always there, however, to scrutinize, or even simply to screen, the contents, for example, of Mircea Eliade's several books on primitive myth and behavior. There, the qualities of homo schizo exude from the time of creation *(illud tempus)* and pattern themselves so as ultimately to reproduce the insane-sane human of today. Whence one may venture among the deeds of the archaic heroes as, for example, in Campbell's accounts, following the gods-driven succession of compulsions, rites, sacrifices, penances, orgies, and aggressions, interlaced with an infantile cute cunning that manifests the earliest pragmatic behavior.

Exemplary in studies of individuals or heroes would be Ulysses or Odysseus[7] whose pragmatic cunning was world-famous, so exceptional was it. He is otherwise a typical survivor of catastrophe; in this case some true disasters of the 8th and 7th century, now well documented, are intended, including the Trojan wars, as well as an echoing of more ancient disaster. Odysseus is an alter ego of the Goddess Athena, a thoroughly dangerous, irresponsible and exploitative psychopath, who never dares to look at himself, an accomplished scoundrel. Ulysses goes into the underworld; he has visions and hallucinations - he is rather paranoid, not only aversive to other people, but pursued by the hostile Poseidon, god of the sea; his reasoning processes are often disordered, when they are not tricky; he is possessed by signs; eternally anxious; homicidal.

Even so, Ulysses was a human with 10,000 years of 'progress' behind him and his story is told by the 'divine' or at least 'highly sublimated' Homer. His life has been faithfully taught to schoolboys by many generations of

---

[7] A. de Grazia, *The Disastrous Love Affair of Moon and Mars,* publ. in xerox, 1968; Princeton: Metron publ., 1983.

teachers, mostly 'normal' and oblivious of this simple and easy interpretation of his character and deeds.

Not even James Joyce saw *Ulysses* in such a light when he wrote his masterpiece by that name; for his hero Bloom is a different kind of schizoid, a "wandering Jew" whose multiple roles were the products of the changes of scene within the city of Dublin and among its people (there being at least two ways of dissociating and cultivating egos - internal movement and external).

The weirdness of the linguistics of free associations found in the novel of Joyce creates a radical contrast to the language of Homer. Homer had to convey a crazy message to the ordinary man, and his language was ordinary; but the leaps and irrelevancies - the great metaphoric stretching - of his style can be seen as chanted liturgy, divine schizoid language, whereas the style of Joyce was ultra-modern schizoid, the liturgy of the individualistic priest of the twentieth century.

When the Greeks and Turks mobilized in a crisis over the Turkish invasion of Cyprus, the author watched young men parading around an island town singing of marching into 'Constantinople' (the Greek name before the city was renamed Istanbul). When he ventured to remark to several by-standers, acquaintances, that with perhaps a quarter of a million deaths Constantinople could be won, they looked at him as if he were a psychopath, or worse, a Turkish sympathizer.

When a contract was let by an office of the U. S. Government in the 1950's to prepare a hypothetical scenario on how to make an unfavorable peace or a surrender in the event of defeat, a public uproar forced its immediate cancellation and apologies from high officials for this insult to *hubris.* But, then, *vox populi, vox dei.*

Theodor Reik tells of how ordinary people, adults and children, and of how prehistoric man, the Biblical Job and Adam, and the Greek heroes were affected by *hubris,* an excessive idea of their competence, a presumptuousness, a belief in the power of their own wishes to transform reality. Prehistoric man must have had an even higher degree of over-estimation of his thoughts and fantasies than modern man.[8] Such would be called delusions of grandeur if met with in the psychiatric clinic. But, says Reik properly, all men have some of it.

J. Jaynes has developed much material on the hallucinatory behavior of the ancient heroes of the Bible, the Homeric epics, and early empires of the Near East. Johnson has done the same, in a less analytic manner, and the present author has concentrated especially upon the psychology of Moses

---

[8] *Myth and Guilt,* New York: Braziller, 1957.

and the Exodus.[9] To Jaynes, the whole of these ancient cultures, perhaps from the dawn of mankind and certainly for the millennia before the eighth century B. C., were bicamerally schizophrenic, the one brain hemisphere cut off functionally from the other, until a loss of faith in dealings with the gods provoked seizures of self-awareness and the beginnings of a complex inner mentation, culminating in the Greek classical age.

Jaynes has identified the greater part of recorded history as a partial recovery of mankind from an early, catastrophically-provoked schizophrenia, and has settled upon the brain-hemisphere split as the locus for the schizoid mental phenomena that we are discussing. In his view, iconoclastic and solitary in contemporary philosophical and psychological discussion, the human mind was behaving 'properly' in what may be recognized as 'the Golden Age of Saturn; ' but then it was sent, even literally, upon the warpath by natural disasters. In our view, the origin of self-consciousness was not in the breakdown of the bicameral mind but in its creation.

Were Jaynes to specify the several disasters, and to allow the original schizophrenia to occur with the very birth of homo sapiens, a remarkable congruence of our theories would result. It is strange - ought I say schizoid? - that in the years he was working on his book he was, to judge both by its inadequacies and by its references, completely out of touch with the vigorous, and even noisy, circle of catastrophists who were working in Princeton Borough, a few hundred yards away, with Immanuel Velikovsky, who was a noted psychiatrist as well as the principal figure in the neo-catastrophist revival.

HELL

The 17th century philosopher John Locke and the 18th century historian-engineer, N. A. Boulanger, both secular investigators, believed that mankind could not have invented the idea of hell unless hell had been an actual experience. These are interesting prologues to C. Jung's concept of archetypes of the mind, where much that governs the unconscious today has been with the human species from its beginnings. Velikovsky and the present author, among others, have presented voluminous evidence for such actual hells, brought on by natural disasters.

Nevertheless, one must consider the possibility that present and historical experiences of hell are part of the self-induced and socially

[9] Julian Jaynes, *The Origins of Consciousness....*, Boston, Houghton Mifflin, 1976; A. de Grazia, *God's Fire: Moses and the Management of Exodus*, Princeton: Metron Publ., 1983.

induced mentation of schizophrenics. Most psychologists believe it probable (perhaps without considering actual prototypical experiences) that the idea of hell is manufactured and processed within the mind. There can be no question that large-scale disasters of burning, earthquakes, explosions, and fall-outs are hellish, and the comment of survivors even of highly localized disasters is frequently 'It was like hell itself.' How did they know so?

Hell may have been sometimes anciently outside of us and affixed its impressions upon us, or it may have been both outside and inside of us since the beginnings. Hercules, the Greek god-hero, who was at least as old as the archaic age, feigned madness. The god Dionysus drove people into collective madness and orgies. Madness has always been akin to divine behavior, and the gods were the producers of hell upon earth. Hercules was identified with planet Mars, Dionysus with planet Venus.

E. R. Dodds, in his brilliant study, *The Greeks and the Irrational,* [10] demonstrates clearly that only very few Greeks of even the classical period, and Socrates and Plato were not among them, thought that man was anything else but irrational and likely to be possessed. Socrates' own treasured second 'voice' is the most famous of hallucinatory companions. "Our greatest blessings come to us by way of madness," he said, "the madness must be of the divine type, produced by a divinely wrought change in our customary social norms."[11] Truth was certainly an ideal, but one to be obtained principally by what we should call today 'occult' processes, involving omens, prophets, oracles, voices, mysteries, ritual, and myth.

Metaphorical or analogical reasoning was paramount, which today we should regard as suggestive but not probative. Deductive reasoning was in the ascendancy, which is essentially 'pulling rabbits from a hat. ' The peculiar kind of empirical induction employed by science *en masse* today was in its infancy with geographers, such as Anaxagoras, and a few others, usually later, such as Archimedes and Thucydides. In all these regards, the intelligentsia was ahead of, but not much ahead of the masses. Socrates was convicted by only a small majority of his fellow citizens, nor was he a very good witness on his own behalf. So we need not venture into 'less-advanced' societies for homo schizo, nor into 'primitive tribes.' He is the hero of historiography.

---

[10] Berkeley: U. of Calif. Press, 1968.

[11] *Phaedrus* 244-a, quoted in Dodds, *op. cit.,* 64.

ORDINARY MAD TIMES

That schizotypicality is the everyday state of historical times seems to be a verifiable proposition. It is unfortunate that in the study of societies, as in the study of individuals, schizotypical and schizophrenic behavior are regarded as departures from a norm, a norm that we can never find. A person is either schizo-typical or nothing. Edward Foulks was hot on the trail when recently he wrote, "Schizophrenia is found world-wide because it has a functional basis in human groups and, until recently, may have provided certain evolutionary advantages,"[12] going on to say that when a society has become stratified and retrograde, schizoid prophets or politicians arise to break down the culture and introduce changes.

They rise and fall - like Jim Jones' American sect that committed mass suicide in Guyana. They are endless in number, in all cultures. The point of distinction is not sanity-insanity but appropriate-inappropriate behavior, or well-adapted-ill-adapted. For every schizoid prophet who is successful, a hundred are crucified.

But that is only a first point. Second, societies have many ways of behaving schizophrenically, ranging from the incorporation of a population in regular wars or killings (the Roman circuses, the Aztec human sacrifices) to the maintenance of a catatonic bureaucracy that employs and stupefies an active population (the thirteenth Dynasty of Egypt, the Confucian mandarins of China, the Soviet agricultural system, the U. S. Department of State). We cannot yet predict if and when a 'stratified and retrograde' society will be busted by schizophrenes. In any event, it is not a question of schizophrene against normal, but of tigers seizing each other's tail, as the Indian children's story goes, and chasing each other so furiously that they collapse finally in a mess of butterfat.

C. Jung speaks of the sudden "disintegration of the personality and the divestment of the ego-complex of its habitual supremacy,"[13] that marks the onset of some acute schizophrenias. It is like experiencing an earthquake, explosions, pistol-shots in the head. These disturbances "appear in projection as earthquakes, cosmic catastrophes, as the fall of the stars, the splitting of the sun, the falling asunder of the moon, the transformation of people into corpses, the freezing of the universe, and so on."

---

[12] Unpubl. xerox mss, 1976, kindly supplied by the author.

[13] *The Psychology of Dementia Praecox.* (1939), Princeton U. Press, 1960, 162.

In 1957, Jung is again conveying the experiencing of schizophrenia, this time of latent schizophrenics, who he guesses must outnumber manifest cases by 10 to one.[14]

> The latent schizophrenic must always reckon with the possibility that his very foundations will give way somewhere, that an irretrievable disintegration will set in, that his ideas and concepts will lose their cohesion and their connection with other spheres of association and the environment. As a result, he feels threatened by an uncontrollable chaos of chance happenings. He stands on treacherous ground, and very often he knows it. The dangerousness of his situation shows itself in terrifying dreams of cosmic catastrophes, of the end of the world and such things, or the ground he stands on begins to heave, the walls bend and bulge, the solid earth turns to water, a storm carries him up into the air, all his relatives are dead...

In the absence of a scientific tradition of quantavolution and catastrophism, Jung, like most other observers, assumes that the individual is displacing his fears upon the religious stories, fairy-tales, and cinema accounts of disaster throughout his life. Of one fact we feel confident: the human mind, whether normal or abnormal, both by past experience and in imagination, has been full of disaster from its creation. Like the main stem of the nervous system, history and historism reaches from past to present. In many schizoid mind stands a Hesiod or a Moses, ready to tell us how it happened *"illo tempore,"* and to transform events into myths, an improvisational and immense creativity deemed a severe form of insanity.

The facts add up to an important bulwark of our thesis, Schizophrenia produces many collective dreams, as well as dreams of personal life. Also, the condition "yields an immense harvest of collective symbols."[15] Some of the collective dreams resemble the Big Dreams found in both mobile and unmoving cultures, of the kind that were reportable to the Areopagus of Athens and the Senate of Rome. We compare these with the output of historism and conclude that in the past, now, and in the future, historism, consciously and unconsciously, is reporting reliably upon the true state of the human mind which is forever being recovered, recycled, reenacted, both personally and collectively, wherever and whenever surrounding circumstances are analogous. Mircea Eliade correctly reports the universal dedication of tribal peoples to the first days of their existence. The continuity of their cultures depends upon celebrating in all major aspects of

---

[14] *Ibid.,* 180-1.

[15] C. Jung, *The Psychogenesis of Mental Diseases,* Princeton U. Press, 1957, 165.

their culture the anniversaries of their birth from chaos and their reception of culture. It takes little comparative analysis to apply fitly the schizophrenic syndrome of mankind to their reliving of the first day. This is their history.

More difficult to propose and accept is our thesis here, that history, as we have known it, since 'the dawn of civilization' is also the return to *illo tempore* by homo schizo in search of his origins. That is, the ceremonial return to *illo tempore* is no more real than the true course of man's history which itself is a form of Freud's compulsive return to the original trauma.

When World War II ended, a psychiatrist, G. B. Chisholm, like many others, was wishing for an end to all the products of insanity such as war. He saw it in a basic psychological distortion that he found in all civilizations. This was "a force which discourages seeing facts, prevents intelligence, teaches mental dissociation and disregard of evidence, produces inferiority, guilt, and fear, makes controlling other people emotionally necessary, encourages prejudice and the inability to understand others."[16] He grasps the symptoms, but persists in superficial meliorism, ascribing the psychopathology of history to bad social policies.

Until psychiatrists, like many sociologists and statesmen, view war as neither inherent in nor an aberration of civilization, but as one way of handling primordial and civilized man's mental and life problems, it is not likely that the war problem can be structured even for preliminary analysis. Earliest mankind probably killed his kind and related kinds promiscuously and in this sense practiced war. Pericot has written that on the various series of pre-neolithic paintings in the Spanish Caves, only one depicts human combat.[17] But we have already argued the prevalence of early violence, and Egyptian murals deal heavily with war.

NAZIS, STALINISTS, AND DEMOCRATS

Often in history, the schizoid becomes schizophrenic, and we see a full clinical disease possessing the collectivity. One of the sharpest episodes of recent memory was the passage of the German nation from a strong self-aware kaiserdom whose schizoid traits were 'lawful' (according to the rules of international misbehavior), underground, and sublimated; to a weak dissociating-ego situation, following a traumatic war, under the Weimar Republic; to the overt schizophrenic state of Nazism; and to a post-World

---

[16] "The Reestablishment of Perceptive Society." IX *Psychiatry* (1946).

[17] In S. L. Washburn, ed., *Social Life of Early Man*, London: Methuen, 1961.

War II republican regime whose therapy was punishment, including self-punition, and an identity with the superpowers of the Age.

The early studies of H. D. Lasswell and F. Schuman on Hitlerism of the 1930's [18] have been supplemented by many more recent works; they employed the method of matching the criteria of clinical madness with the speeches and writings of Nazis, the characters of the leaders, their actions and public policies, and the response to these of German public opinion. Their conclusions are typified by this sentence from Lasswell's study: "The conscience for which [Hitler] stands is full of obsessional doubts, repetitive affirmation, resounding negations, and stern compulsions." Identification, displacement and projection, obsession, repetition, negativism, aversion and compulsions nest in this one sentence, most of what composes human nature in fact.

We note, concerning the effective Hitler appeals, the logic of metaphor, the profligate use of analogies, the 'reasoning by right brain,' his 'effeminate intuition,' his artistic background, and we raise a question for those who feel that the 'left brain digital logic' is somehow more at fault for violence than the 'humanist' right brain of the poet and musician.

We are reminded of a case of Bleuler. "A catatonic notified the court that his illness had been diagnosed as paranoia and the apparitions as hallucinations. 'Be that as it may, ' the patient asserted, 'there are still sufficient reasons to proceed against the gang.'" [19] Then we have Hitler personally conducting the massacre of his own men, many of them loyal, in the infamous purge of June 30, 1934, asserting to the German people thereafter that, granted these men may have been innocent, they still deserved to be killed because (as far as one can disentangle his words) they were guilty of making him suspicious and this was the same as threatening to destroy Germany. These alleged conspirators, agitators and destroyers were "poisoners of the wellsprings of German public opinion," a metaphor suitable also for arousing deep feminine sexual fears of impurity and impregnation, and of antisemitism, whose folklore had utilized the same allegation against the Jews since time immemorial.

To stress the metaphorical logic, we recite, too, *Mein Kampf,* where Hitler had written, "All great movements are movements of the people, are volcanic eruptions of human passions and spiritual sensations, stirred either by the cruel Goddess of Misery or by the torch of the word thrown into the

---

[18] Lasswell, "Psychology of Hitlerism," in *The Analysis of Political behavior,* New York; Oxford U. Press, 1947. Frederick Schuman, *The Nazi Dictatorship,* New York: Knopf, 1935.

[19] Eugen Bleuler, *Dementia Praecox, or the Group of Schizophrenia,* 1911, J. Zinkin, tr., N. Y.: Int'l U. Press, 1950, 128.

masses, and are not the lemonade outpourings of aestheticizing literati and drawing-room heroes."[20]

To generalize about history cannot be scientific, and, if scientific, cannot assemble its volumes of proof, and, if it can, it will certainly be misinterpreted. Can we agree that all history is not Nazi? Of course, but how much of history is such is a matter to report as well. If the Nazis had not deliberately put to death millions of Jews and other human beings, a German history of fifty years would not be studied as a case of collective madness. Yet it would still have been history as a recital of schizotypicality.

The Stalinists of the U. S. S. R. and its satellites largely evaded the stigma of madness; they committed millions of murders, assigning as pretexts collective mutiny as with the Kulaks and Cossacks, or military necessity as with the massacre of the Polish officer corps at the Katyn Forest, or of conspiracy against the worker's state as in the 1930's treason trials and purges that brought death to many thousands and filled the deadly concentration camps of Siberia.

Rigidity frequently takes the form of logic and principle as a feature of German character; it is a quality that will not compromise with politics or with 'how other people feel.' This heavy schizoid trait is better camouflaged by acceptable doctrines in other nations. The Russian behavior was, for instance, generally believed to be "more human," partly because of the humanistic ideology and "underdog" connotations of marxism. When Marshal Petain, later to be condemned as a traitor, became the hero of France in World War I, it was because he took the sternest measures to insure that a million or more Frenchmen should be killed or wounded in the Battles of Verdun. The German military leaders were equally distinguished at Verdun. The American and British destruction of enemy cities in World War II were justified as combination of retaliation and military necessity. The list of such behaviors is exceedingly long and falls back to the dawn of history - to the glee of pharaohs inscribing on their temples and tomb walls what armies of men they slew, what slaves they took, what towns they destroyed, what loot they carried home; it retires also to the Vedas of India; to the epics of Homer; to the Books of Moses and Joshua.

Yet it would be a cheap trick to let the case for homo schizo in history rest upon war and civil violence. We should appreciate that man is at war only half the time. Perhaps no more than a fifth of all deaths since humanity began have been from violence, directly or indirectly. The shadow of war and violence is always over mankind, of course, and this shadow uses the visions and rhetoric of insanity; and the history that is told is largely the

---

[20] From Houston Peterson, ed., *Great Speeches*, N. Y.: Simon and Schuster, 1965, 757, 759.

stories of war, written by the greatest number of historians, and coursing through the historical senses of people *en masse.*

We can proceed beyond warfare to larger realms; not a trick, but as real as can be, is the claim that collective behavior can have the same psychological adjectives applied to it as individual behavior. One needs to be careful. There is no brain, heart, liver, or limbs or phallus, etc. of society except as metaphor. As the American marine general argued, when told that his national policy was to win the hearts and minds of the Vietnamese, "Grab them by the balls, and their hearts and minds will come along with them."

But if one holds to a quantitative mode of thought and discourse, one can say: "there are 2, 3, 5, 10 out of 11 persons employing the same mechanisms," and even extend this to such aggregates as the "French governing group," or even "the French people," or "a typical French practice," or "a change in German attitude" meaning over x% changed, etc. This empirical and quantitative mode of thought must be emphasized, lest, on the one hand, strange and confusing metaphors be employed and trusted - like "the heart of the nation" - or, on the other hand, lest it be regarded as unscientific to speak, for example, of a collectivity of persons being traumatized by a catastrophe.

What happens to a 'group' happens to the individuals composing x% of the group, or it does not happen to the group at all. When we say 'a group is schizoid, ' we mean that the traits of human nature are all operative in varying forms among the group members in subjective, interpersonal and external transactions. When we say that this group becomes Nazi, we mean that its ruling element and a significant portion if its members are acquiring a preponderance of Nazi attitudes and exhibiting Nazi behaviors. We can also say that a group depersonalizes, as Germany did following World War I, developing, as Sebastian de Grazia has described it, an acute anomie, not being able to find itself or an appropriate image of itself.[21] The United States appears to have gone into such a state with the first generations to follow World War II. We await a masterly treatise along these lines, but meanwhile are diverted daily by episodes such as one momentarily in the news as these lines are written, of a fourteen year old boy who raped and murdered his girl-friend and who exhibited her unburied corpse from day to day to a dozen acquaintances, none of whom called the police. This is acute anomic behavior.

However, any historic (i.e. past) behavior, whether selected randomly or chosen as an extreme test of the proposition, will exhibit the full range of

---

[21] *The Political Community,* Chicago: U. of Chicago Press, 1948.

schizotypicality, because that is the only way that people could ever behave, as they can only behave now. Some extraordinary incidents are chosen for their atypicality, and these compose most historiography and 'news.' Schoolchildren read that Abraham Lincoln walked miles to repay a few pennies to a lady who had been given the wrong change in his store. Psychiatrists such as Clarke have dwelt upon this incident, in analyzing Lincoln's character.

But let us say that Lincoln gave his next customer the proper change. Isn't this a simple transaction, a bargain, a sale? What would be psychopathological about *this* ordinary transaction? Of course, firstly it is assumed history and not taught - why? Because it lacked significance, significance meaning something sinister and obviously schizoid, whether positively moral or evil. Or because it would make dull reading, which is proof once removed of the same. The total setting, the total action frame of historiography is human and schizoid.

Deviancy, terror, violence, and pornography must constitute most of all that has emerged as literature, art, and history. A great many routine actions bear the stamp of rationality simply because they are conducted in an accepted cultural structure. A machine-gunner, who kills twenty men whom he has never met, is simply making small change: there is nothing psychopathological about him. There stands a fine monument to the Machine Gunners of World War I alongside Hyde Park, in London; the inscription on it reads from the Bible, "Saul slew his thousands, and David his ten thousands." (I Kings 18: 8)

Let me clarify by means of another case. In *Psychopathology and Politics* (1930) Lasswell speaks of the man who hates his father and tries to kill the king, and accords to such behavior a formula: that the political man (terrorist) displaces private motives (father hatred) onto public objects (king) and rationalizes it in terms of the public advantage (tyrannicide or republicanism). This sounds pathological, abnormal, and rare, but a few moments of analysis will reveal that everyone engages in precisely the same mental operations and activities in everyday life. That is, one aggregates 'private' and 'public' objects by displacements, and acts, in one way or another, similarly (so far as concerns his motivation) with regard to both types of objects.

RELIGION AS CUSTODIAN OF FEAR

What else has man done other than prepare for and engage in conflicts and war? He has practiced religion as much or more of the time, and it would be well if one might present a concise statistical inventory of all that has gone on in the name of religion. Without such a summary resource, and not wishing to recapitulate the extensive exposés by eighteenth and nineteenth century writers such as Voltaire, Boulanger, Feuerbach and Frazer, and especially because a more systematic analysis of religion is intended in a later volume of this series, our remarks here must be brief and linked closely to our theory.

I have elsewhere cited the ancient realization, expressed in the saying of Lucretius, Statius, and others that "First of all the gods created fear," and "First of all fear created the gods." Fear is in all life but especially in mankind. Man, upon his first full appearance, created his gods to be responsible for his fear; moreover, they were created for the major purpose of controlling his fear.

If this be so, it should not surprise anyone that, since the first days of human history, religion has been the principal custodian of all the major aspects of fear. And that the general fear is incorporated in the routines of life and any particular fears that arise are invariably fitted to religious fear before they are released for testing in more pragmatic areas of life. Divine action has been the first hypothesis for explaining every event.

Continuity is perceived as pursuance of divine behavior and teachings; change is seen as a violation of or an instruction of the divine. The divine is the creator and the mediator of all things, the intervening variable between cause and consequence that is too often denied or left out by those ancients with hubris and those moderns with science.

Human behavior is continuity: will not most readers agree that religion suffuses all that is long-enduring, routinely undertaken, and traditional? And change is always a rebellion against some aspect of some religion usually in the name of another aspect of the religion or of another religion. That Marxism, a non-religious doctrine of social science, is practiced without a heaven and invisible god is apparent wherever it prevails, and often it rests on top of a population retaining its traditional religious affinities.

The Chinese have for millennia been fond of what we have called "ritual counting;" and when the youth of China was given a *carte blanche* by Mao Ze Dong in 1967 to tear down traditional institutions, including the covert religious practices of Confucianism, they were told to destroy "the four olds," old thought, old culture, old customs, and old habits. The terrible aftermath was referred to popularly as "The Revenge of T'ien," the ancient living Heaven.

That religion is everywhere schizotypical is not difficult to prove, if a hypothetico-empirical science is assumed, i.e. a science based upon the tenuousness of propositions and the rules of material evidence, for this arrives speedily at Thomas Hobbes' assertion "The fear of things invisible is the Natural seed of Religion." That it is abundantly schizophrenic in the usual definition of disease, making of its practitioners either outright schizophrenes or followers of the same, also emerges from a simple and fair reading of the religious record in history. Delusions, hallucinations, paranoia, and anhedonia are at the core of every great religion and tribal sect. It is ironic in the extreme for devotees of religion to explain the madman, i.e., the assertedly schizophrenic, as one who has fallen away from religion and is therefore accursed.

Atheism abandons celestialism and anthropomorphism, but cannot, all hopes to the contrary notwithstanding, divorce itself from schizotypicality. Elsewhere, I have ventured to say that real celestial activity was the original sponsor of religion; and it has always been a reinforcer of traditional religiosity, as people view the skies once again as part of their religion. Catastrophes are breeders of typical religiosity. The seeds of many memorial generations lie dormant, awaiting the occasion of disaster to sprout.

But in the interim of calm skies, a few people become atheists and claim a capacity to think for themselves, to think in hypothetico-empirical terms, and to act pragmatically. Modern western culture is even dominated to some extent by atheistic thought. Still there is no question of a basic change in humanity occurring. The same mechanisms and processes of perception, cognition, decision and action occur, only without an area of displacements hitherto filled with the Heavenly Hosts.

An atheistic bookkeeper in a soviet machine factory, bred of three generations of urban atheists, fills his mind with identification with the dead Lenin and heroes of the communist movement: he projects his wishes into the leaders of the Soviet Union, ascribes to his boss and American imperialism feelings of hostility toward him that he feels toward them, finds solace in work and in alcohol, is scrupulous and neat to a fault, nurtures a constant cold in the head, plays psychological games with his family and neighbors, and so on. "He is a good man," they say of him and he will be buried in the earth of Mother Russia without benefit of clergy. All of which is to say that this man is of the ilk of the friar of a Byzantine monastery that once stood next to his cemetery.

Celestialism, sky-religion, which has marked the history of worship, was a contributing factor to the creation of homo schizo and primordially paramount in the filling of his mind with displacements and ideas, but man does not require a continuous experience of sky activity, nor a conscious belief in its historical or present actuality, to be either mundanely religious or atheist. However, in no case can he cast off the schizotypicality that accompanies celestial religion by becoming either mundane or atheist.

"Cambia il maestro di cappella, ma la musica è sempre quella!" - the choirmaster may be changed, but the music is always the same.

UTOPIANISM

A final escape is solicited, not historical, but utopian. "Imagine a group living communally in houses of a settlements that they have built. There they grow food and animals and eat them, and fashion tools to make necessities such as clothing and furnishings. They cure disease empirically and save only enough for a rainy day. They love their children and old people and live in peace with their neighbors. They profess no religion."

Now perhaps this community has never existed. But, if it did, would not its people be called truly homo sapiens sapiens?

No. These people are apparently schizotypical. The very conception of them and the conception they have of themselves - the utopia - is schizoid. The utopia begs all questions of its creation and leaves us with dogmas of conduct and consequence. How they positioned themselves for the utopia is unknown. Like all utopias it is an exercise in the omnipotence of thought: to think of something is to create it.

Yet, passing over the absurdity, examine the activities of the community. All require severe consensus. How shall decisions be made, by what system of voting? What plan will be devised that is not descended from the prehistoric pillars of heaven, north-south orientation, the planetary circuitry of the walls of ancient Babylonia, the star of Saturn? What shall the diet - that Pandora's box of phobias and compulsions - consist of? What is to be traded for the tools, or will we be here in an autarchic Stone Age village? What identifications are to exist between commune and neighbors? What language will be employed to deal with them: are they *vous* or *tu*? Will the diseases be all organic, to avoid problems of definition, and will psychosomatic illness be denied or absent? Can this last denial work, considering the rigorous training imparted to the children, who must, despite this heavy discipline, love themselves, their old people, and their neighbors. Much must be set ahead and back in time, too, for to love the old means to respect the olden times that the old like to talk about. Planning is everywhere: the crops to be harvested, the goods to be made, the curricula of teachings: the saving (how much?) for the rainy day (when will it next rain?). Practically all psychologists (except perhaps one such as B. F. Skinner who has written of such a place that he calls *Walden II)* will see in this mythical community a highly integrated and coordinated set of schizotypical human behaviors. They will foresee in it a propensity to totalitarianism and religious revival, once a disastrous threat appears from 'the friendly neighbors,' or 'benign nature,' or 'traveling in foreign places.' The community is ahistorical, an impossibility. It is founded upon a non-existent kind of human nature. Should it "succeed" in any other sense, it will succeed as a grand delusion.

DARWINIAN HISTORISM

It may not be long before there is a general realization that the foundations of Charles Darwin's idea of the origin of species (1844) and the descent of man (1871) were intellectually weak, and that the success of Darwinism was, like that of Alexander the Great and Isaac Newton and Napoleon Bonaparte and Karl Marx and Sigmund Freud and Albert Einstein, first of all a success of the opinion thunderstorms of the times.

To repeat a theme of the first chapter of this book, Charles Darwin argued often on a *post hoc ergo propter hoc* basis: where organic variation existed, it must have been preceded by something less advantageous, and what brought about the change would be called "natural selection." Natural selection was more than a name to him; it was a reality, even a dogma. Influenced explicitly by Lyell who saw long, uniformitarian processes of change in the rocks of the earth, and inspired by Malthus who saw famine, war and disease as always ready to cut down a surplus population to viable proportions, Darwin could examine one form of instinctive behavior after another in animals and purport to find in their variations "consequences of one general law leading to the advancement of all organic beings, - namely, multiply, vary, let the strongest live and the weakest die."[22]

This was the principle of "survival of the fittest," wrote H. Spencer, in an approving vein. A mutual approbation society grew up among economists and biologists. Its cold, dogmatic line of thought provided the largest, deepest source of aggressive *laissez-faire* competitiveness. To its influence, many commentators have ascribed the breakdown of the human mind in the last century - meaning the open exposure of the schizophrenia of human nature in cultures. That is, many sociologists, anthropologists, literary critics, and philosophers agree: the historism of Darwin did not settle the minds of homo schizo. He hastened the break-up of the selves system of his age.

Thus can we say that Darwin, as an historian, for that he certainly was, would have been unconsciously seeking, according to our own theory, to

---

[22] *Origin of Species*, 1859, ed. 1936, 208.

provide his clients with the means of controlling their ever-anxious schizoid minds. The argument is surprisingly simple, and even well-known. The minds of his clientele, the *cognoscenti* and *literati,* and the radical and socialist revolutionaries like Engels, were already in a distraught condition; for they were rejecting mosaism in religion and feudalism in politics. They were desperately agitated and impatient.

Darwinism provided a new swarm of displacements, a set of obsessive problems, an outlook for aggression against well-defined authorities, even a stable primate-mind that could view remorselessly the gradually changing social scene of nature. Darwin himself probably saw his mission, and, if his personal despair is significant, realized he had failed to accomplish it.

Despite all that has been written about him and the history of biology, much more could be said than this study can comfortably bear. I can only try to oblige Darwin's requirement, expressed in a letter to Thomas Huxley: "It would take a great deal more evidence to make me admit that forms have often changed *per saltum.*" [23] I would probably suggest, too, a good psychiatrist. The "dreadful but quiet war of organic beings going on (in) the peaceful woods and smiling fields," as he put the struggle for survival, was going on in this abnormally intellectual specimen of homo schizo as well.

I think that, personally and as a typical man of his times, he could bear the infinite trench warfare of his theory more than the bombastic war of catastrophism, implied in the euphemistic word *saltum,* the leap. Catastrophism was the world of the Old Testament and of his father's and wife's character. He was the invalid fighting the point-by-point, day-by-day war all of his life.

On the eve of the publication of *The Origin of Species,* he wrote his cousin that "I have been extra bad of late, with the old severe vomiting rather often and much distressing swimming of the head. My abstract [of the manuscript on which the book was based] is the cause, I believe, of the main part of the ills to which my flesh is heir to..."[24] Although he had conquered conscious mental revulsion against his theories, he could not suppress the psychosomatic revolt. He was a gentle man, people agree, but with a compulsion to speak out rebelliously and aggressively in displaced intellectual forms. His world of nature was his world of struggling selves within.

I also wish that I might do a proper analysis of evolutionary theory in biology and anthropology, especially as it concerns man. Instead, I can only guarantee it to be still a happy hunting ground for the logician who is

---

[23] *Life and Letters,* II (1860) 274.

[24] Ralph Colp, *To be an Invalid: The Life of Charles Darwin,* Chicago: U of C. Press, (1977).

biologically trained. It abounds in evasions, question-begging, circular arguments, *ex post facto* 'discoveries,' *post hoc ergo propter hoc* reasoning, proof by selective example, *hysteron proteron,* and doctrinaire assertions. This is all aside from the paucity of evidence on important issues, the failure to recognize important issues, and the entanglement in trivial research.

The final nemesis is the ever-present, ever-available two-way switch between the genetic pool and natural selection. Natural selection can never fail as the means of evolution because it will always presumptively find among the genes of any species whatever precise gene, unknown of course, is needed to explain a given step in evolution. With this suppositious entity, any hole in the hulk of natural selection can be plugged. This is gene 'Q,' the potential quirk that conveniently enters the gene pool prior to whenever the time arrives for it to be called forth, potentiality into actuality.

If these allegations are deemed too severe, I may at least hope that biologists, psychologists, and anthropologists will realize in them a suggestion of the need for an explicit union of social, psychological, and biological theory. Also, in the course of writing this volume, it became clear to me that I was presenting neo-darwinists with a strong and original argument for their case, even while establishing my own case. This has occurred by locating and simplifying the instinct-delay mechanism as the force of transition from primate to man. It is a factor that is to a high degree quantitative and therefore could be considered capable of sustaining many minute changes by mutation and adaptation over long periods of time. I hope that this bonus will compensate my critics for reading what otherwise may have appeared to be an offensive attack upon "the true facts" of evolution and culture theory.

CHAPTER VIII

THE HOPEFUL MONSTER

My story of the "hopeful monster" is nearing an end. Given the physique of an ape and a troublesome miniature computer, he has harried the whole Earth, scuffed about on the Moon, and can sing "Aida." I see no signs of the angelic in his origins and history, only in his delusions and pretensions - but what can one expect from a schizoid?

Nor have I discovered substantial grounds for any theory of the origins of human nature except that of homo schizo. Aside from special issues and errors of fact, none of which, it may be hoped, are fatal, four major criticisms can be aimed at the theory. These can be stated as follows: first, that the catastrophes of the human creation scenario did not occur; second, that humanization was not a hologenesis but a process occurring point-by-point; third, that the human species appeared much earlier than 13,000 or even 50,000 years ago; and, fourth, that mankind is not genetically schizotypical.

Any one of these criticisms can be offered by itself, while with-holding the others. Moreover, all four of them can be advanced together. At the same time, I can be correct in any of these four regards and also in all four of them, to wit, the catastrophes, the hologenesis, the recency of humanization, and the schizotypicality of human nature.

REAL AND PSYCHIC DISASTER

On the issue of catastrophism, I should repeat the thesis. There are three kinds of disastrous intervention in the process of humanization: First, catastrophes are invoked as requirements for mutations, biosphere destruction, and atmospheric transformation, without which the human species -- and many others -- would be most unlikely to evolve. I have sketched the evidence and the character of such disasters and shown how they would enter into the quantavolution of mankind.

Second, the metaphor of catastrophism is applicable to the hominid mind as it was destroyed and the human mind composed. In this sense, the human came about as a schizophrenic psychological disaster. To sharpen the point, one can imagine that a team of scientists, knowing much more than we do now, and expert with drugs and surgical instruments and experimental environments, might convert a hominid mind into a human one. The team could then announce, metaphorically, "We have today fashioned a fearful, power-seeking, disaster-prone maniac interested in everything in the world."

The third form in which catastrophe intervenes is once more in the non-metaphorical mode. Natural catastrophes occurred after humanization on a grand scale and at intervals of time. These provoked and reinforced the catastrophic character of the human mind, entering, with 'unfortunate' compatibility, upon the interior and transpersonal melee of individual and collective psychology.

It would be unwise to place the burden of proving the true natural catastrophes entirely upon this one book. At the least I may refer to my books of the Quantavolution Series for more theory and evidence and then to the classics and rapidly growing literature of quantavolution, as cited in those books.

The human mind - an itself that perceives itself as a disaster emergency - is sandwiched between natural catastrophes that preceded it and natural catastrophes that succeeded it. We are not surprised, given the catastrophic interfaces, that the human is often an unreliable observer. Still, although the world is ultimately to the mind a coded set of illusions, and although this mind must always possess a great many delusions about these illusions, there does exist a sense of reality, a pragmatic mechanism simulating the animal's instinct to respond to a stimulus.

The pragmatic mechanism permits humans to distinguish between more or less delusionism. Legend, myth, history, and thought can be reality-tested for their degree of 'excess delusionism. ' The mind of homo schizo, in sensing the outer world, can build a battery of tests to discriminate more disastrous from less disastrous conditions.

A RECENT SMALL SHARP CHANGE

Whatever one's ultimate judgement on the issue of catastrophism, a second criticism may be leveled against the general theory of homo schizo, namely, that humanization was a point-by-point process, not a hologenesis. In this regard, one can review the psychological theory of this book and of its companion volume, going beyond the evidence, which is thus far scarce, concerning both quantavolutionary and evolutionary theories. Our theory here says: the critical change in the pre-human creature was probably small, a significant depression of instinct-response speed, but the effects were an avalanche, pouring into all aspects of behavior, internal and external, and prompting an immediate culture.

Our position, disregarding the evidence momentarily, is logical and theoretical; it is well illustrated in the scenario of the simple club-carrying creature: he has to be a fully human person. No matter by what door one enters into human behavior, one enters upon the domicile of the human being. The central nervous system, mentation, and culture are holistic -- all must be related to all.

Whereupon a third criticism is ventured, that the human species is known to be very old, even though human behavior and culture are not demonstrable until the Upper Paleolithic age. This criticism usually is brought in to support the second criticism, but is logically independent: one may have rapid-fire point-by-point evolution occurring in a short time. The question here is how long ago did humanization occur. My position is that the time scales are grossly distorted, that 'three-million-year- old-hominid bones' are perhaps no more than thirteen thousand years old. I have indicated signs of weakness in the tests of time upon which so much faith is placed.

Either we are dealing with a hominid who is humanly incapacitated, or we must drastically shorten the time scales. A human who is distinguishable three to five million years ago, who then disappears for three million years, and who then emerges with a culture, is a contradiction in terms; he was not human. But could he have become clandestinely a human only 100,000 or

200,000 years ago? If one has to speculate in this fashion, then the debate will become a free-for-all, no holds barred.

THE UNREDEEMABLE APEMAN

The fourth and last major objection to the theory of homo schizo is this: however he came about, man was born a rational animal, in whom signs of schizotypicality are abnormal, even if frequent. If someone will argue along this line in the face of the abundant evidence and citations advanced in this book, not to mention its companion volume, then that person is ready to accept the implantation of a soul by intelligent beings from outer space, or by a god of his choice, as the critical step in human genesis.

Ironically, the ethologists are on the proper track: man in his 'rational' nature is most like an ape, seeking the simplest solutions and fastest decisions that a brokendown instinct apparatus will allow. As Blaise Pascal said, long ago, "There is no man who differs more from another than he does from himself at another time." It is in one's erratic attitudes or values, in one's conflicts, and in one's unsatiable curiosity that a person is most human. No other animal species is so ineluctably schizotypical. The so-called irrational element of people is therefore their authentically 'normal' constitution.

Surprisingly little systematic scientific theory of the genesis of human nature exists. We know rather well, in this age, what constitutes a general scientific theory, and if theories of human origins are scarce and defective it may be because their empirical foundations are absent. In such circumstances, we can offer the theory of homo schizo with greater confidence, realizing that it is not just 'another theory.' Let us repeat then the theory, in summary form:

Given the conditions that must have attended human creation, human nature must have been of necessity schizoid. Furthermore, judging from what is known of his early behavior, culture, and history, he was in fact schizoid.

As to the first point - what 'must' have created a schizoid human in the process of nature - we allude to the constitution of the primate, from which man derives so many mental and physical attributes. The fascination that crowds humans into the monkey house of the zoo reflects the intuitive recognition of similar species. An unending stream of detailed studies just reinforces the resemblances of physiology, anatomy, and behavior. But reflectiveness, symbolism and reasoning on widely displaced subjects are missing.

The forces that generate species by mutation are constrained by the necessity to work on what is already potential, in order for the species to survive in a physical, as well as environmental sense. Mutations are not purely random, not quite blind, but strike the target like poor archers, off center. In the case of humanization, the key mutation produced directly or indirectly a fatal indecisiveness, whose first outward evidence would be a crazy, that is, misbehaving, hominid. The preconditions for mutation included natural particle or viral storms of sufficient scope, intensity, and duration to cause a great many mutations.

To fix by conventional chronology a certain date for the birth of mankind is risky and might mislead; the compression that we force upon the usual timetables reduces millions of years to mere hundreds. The boundary line between some pair of ages that run from the Cretaceous to the Holocene (a sixty-million year interval in conventional geochronology) might have witnessed the first humans. On such occasions the skies changed, the atmosphere was turbulent, the Earth was deluged, the lithosphere was refashioned, and great numbers of species were extincted. Thereupon might the human, whether by mutation or radical adaptation, have originated.

In themselves, the changes from hominid to human may have been anatomically negligible. Even the swollen cerebrum is scarcely distinctive. Hence what happens inside the brain is all-important, for that is what translates into uniquely human mentation and behavior. We have argued that what happened had to happen at once, in a hologenesis of mind and culture. We have demonstrated that little time was needed to permit the speciation of man and that probably little time was actually available, the current geochronometry notwithstanding.

Further, the speciated man was genetically predisposed to culture. Culture was inevitably and promptly determined by the human quantavolution. Recognition of this process has been blocked by the same evolutionary and uniformitarian ideology that insists upon point-by-point speciation; point-by-point cultural evolution is impossible. Culture is species specific behavior of homo schizo. He finds culture as he finds a water hole or a mate. And this culture is a monstrosity of nature, whose very existence proves that man is the only species that dwells outside of itself, out of its mind.

SCHIZOTYPICALITY AND HOMO SAPIENS

The primate ancestry, the turbulence of the environment during the birth throes of the species, and the basic human culture all point toward a creature who is perennially distressed from having to invent his own mind. He had now to behave in the pattern of what is today called schizophrenia.

He was depersonalized. He constructed a multiple personality and operated under a confederational ego. He was fearful and anxious continuously and without sufficient cause. He has remembered a terrible past which, inconsistently, he has forgotten and displaced.

He had an unceasing and unbounded need for control of himself and pursued all semblances, fakes, illusions and self-deceptions that seemed to give such control. He displaced madly. He has always thought by displaced association and projection. He animated nature. He sought the eternal. He had immediately to establish the trappings and rituals of culture. He suffered religious delusions and made and unmade gods, under the illusion that these gods were busily making and unmaking him. He killed and ate his kind.

He was obsessive and compulsive. He has consistently disliked himself and others, and has been characterized by aversiveness to people and ambivalence. His ambivalence extended to himself, to others, to the gods, to all of nature. He has loved and destroyed all of these.

Nor has he been happy. Anhedonia, the 'joy' of suffering, has always been a major human trait, though often so deeply buried in his culture that he can go about 'happily' denying its presence. He has been typically paranoiac; never could he manage to build more than a narrow crust of trust, even though paranoia unleashed the most self-destructive kinds of behavior.

He symbolized internally and then extruded portions of his code for external communication. The symbolist process expanded enormously to take in the whole of his world and of nature. Everything perceived and conceived received its code name. The linguistic process was done not once and for all, but repeatedly, thousands of times, in thousands of cultures and at different periods of the culture. He made fetishes of signs, words, and symbols.

He has always envisioned a future, but the future squeezed in and out like an accordion, from the next beat to forever, dragging along a ragged melody, out of time with everyday behavior and history, and changing its tune from one moment to the next. He could add and subtract, which from time to time amounted to marvelous intricacies of mathematics and logic, yet these formed always a limited, not powerful, element of his character.

He designed and valued decision in many forms as the substitute for the instinctive behavior that he lost and would dearly love to relocate. Great mythologies and sciences of decision emerged. He ventured into totally 'unproductive' fantastic and philosophic contemplation. For dreaming so much while asleep and awake, he is uniquely distinguished among all creatures. In the beginning, as ever, he became mythographer and historian, the schizoid recorder of his own schizotypicality.

To conclude, we have found no symptom of mental illness, or schizophrenia as that is broadly construed, which has not been an important and 'normal' part of human nature from its inception. We find, also, that there has been no large general and persistent pattern of human thought and behavior that cannot be subsumed under the symptomology of schizophrenia.

The name, homo sapiens, and especially homo sapiens sapiens, given to him, is a misnomer and he should more accurately be called homo sapiens schizotypus. Indeed, there may turn out to be, by tests refined beyond those that are presently validated, a mixture of human natures, including hominidal forms that cannot survive or regenerate as humans without instant heavy administrations of culture, and then several others, one or more of whom was responsible for the invention of culture and the great changes of history. Possibly there may be, among the latter, some genetical structure that is so fully cerebral and ego-controlled that it can be called "sapiens sapiens."

It is, after all, mainly a convention that bids us call all people by the same species name. There is no natural law that demands that all people be of the same species in that they apparently can interbreed. Even at that, some humans can interbreed only under high risk circumstances, at some risk to themselves and their progeny, owing to genetic differences.

We may also suspect that the stress on species intra-reproducibility may be an offshoot of the peculiar sex-sublimated English nineteenth century environment, in which Darwin and his friends were heavily immersed and in which animal breeding was of large interest, as Darwin himself evidenced, and in which 'good breeding' and genealogies within human groups took on a sacred aura. By contrast, many peoples of the world, including Greeks and Hindus, have claimed that the human soul could migrate, in all or in part, hither and yon in the animal kingdom, transcending zygotic barriers.

Genetic differences among individuals become minor or major by definition, by public policy, by fiat, one may say. While ordinary people, societies, scientists and the intelligentsia have their eyes upon certain visible differences of culture, upon skin color, stature, and certain other differences, not apparent but tangible, such as blood groupings, other more basic behavioral and mental differences may stand unattended. These latter may appear to be so important, when ultimately perceived, that they will erase not only the traditional insistence that all people are of one species but also the thesis of this book that all people are to be presumed schizoid. More likely to occur would be the uncovering of genetic differences that are too "minor" to suggest drastic eugenics.

The worst possible occurrence, which would at the same time be the best possible, would be the discovery of a truly homo sapiens sapiens among the population in 'pure' or 'diluted' form. Even to be able to recognize scientifically such a type would be difficult if not impossible, since we should have to recognize something that we are not, to lift ourselves by our own bootstraps, so to speak.

If it were recognizable, presumably it would be a person whose self-awareness is infinite but at the command of a calm will of a solid genetic ego. It would be a person who can displace without anxiety, but who also has a will to displace infinitely. This person would make and unmake habits with only instrumental motives in mind, could believe in causes without bias and prejudice, could emotionalize warmly without commitment to irrelevant choices, utilize large portions of his or her cerebrum for calculations according to a presently unknown logic, control his nervous system and physiology at will (that is, by perfect psychosomatism), be prudent but fearless - in short do many things naturally that we have here come to believe cannot be done without contradicting nature.

Such would be an 'ideal' species, as we would megalomaniacally describe one, without reference to the reality of homo schizo. We construct this ideal, in fact, with as little appreciation of its possibilities and likely genetic mechanisms as we imagine the 'intelligent beings from outer space, ' with only the vaguest ideas of how such beings might be anatomically and behaviorally designed. There is, I would conclude, only a rare chance that such a species exists among us and, if it did, could be found, and, if found, could be eugenically engineered and maneuvered into commanding position in the development of a real homo sapiens sapiens.

Most probably we are confined to homo schizo in ourselves and in society. Our tactics, and to some unknown degree our eugenic policies, must be to trick ourselves and others into certain ways of behavior whose consequences we desire and accept. These tricks can carry us a certain distance towards utopia.

In addition, as we discover the right tricks (I realize that I should be calling them applied social science or humanistics), we can be alert to discover certain quantitative genetic differences that reliably distinguish those human schizoid constitutions that prefer our tricks -- our solutions -- and are docile respecting them. These can be genetically assisted.

So a cultural and genetic kit-bag may eventuate that will give us a new typical homo schizo, ideal in these senses rather than in the unrealizable megalomaniac conception that was hypothetically formulated above, which, incidentally, resembles the far-flung schizotypical visions of man that are commonly voiced by philosophers and politicians.

About the author

Alfred de Grazia was born in Chicago on December 29th, 1919. His father was a musician and conductor. After receiving his doctorate from the University of Chicago in 1948, he taught social theory, political psychology and behavior, and social invention for some years at University of Minnesota, Brown University, Stanford University, and finally New York University, lecturing widely in establishments of higher learning at home and abroad. His works on politics and theory helped to establish scientific method in the field of political science. He worked at military psychological operations in three wars. In World War II, he was affected to OSS in the first year of its inception and fought in seven campaigns in North Africa, Italy, France and Germany. A lifelong occupation with social invention was reflected in the founding of the magazine *The American Behavioral Scientist,* the design of information systems, and in public policy studies that began with the First Hoover Commission on the reorganization of the federal government and continued through the publication of over a dozen books, and participation in numerous commissions and panels. In 1964, he published *The Velikovsky Affair.* From 1963 he began a new series of studies on ancient catastrophes and their effects upon natural and human history, which eventuated in the Quantavolution Series. He also published poetry, plays, and several autobiographical works. He lives presently in France and in Greece, with his wife, French writer Anne-Marie de Grazia. His brother Sebastian was a political scientist, laureate of the Pulitzer Prize. His brother Edward a prominent First Amendment lawyer, his brother Victor was Deputy-Governor of Illinois. His daughter Victoria is a historian and member of the American Academy.

His two million+ visitors/year website: http://www.grazian-archive.com

See also: Q-MAG.org - the magazine of Quantavolution: http://www.q-mag.org

LATEST PUBLICATIONS *(all Metron Publications)*

40 Stases & Theses: What Is To Be Done With Our World?

Gods Fire: Moses and the Management of Exodus

Cosmic Heretics: A Personal History of Attempts to Establish and Resist Theories of Quantavolution and Catastrophe in the Natural and Humans Sciences, 1963 to 1983.

America's History Retold (Vol. One): Conquest, Colonialism and Constitutions (2012)

America's History Retold (Vol. Two): Originating American Ways of Living and Working (2012)

America's History Retold (Vol. Three): Shaping Earth's Cultures and Powers (2012)

A Taste of War: Soldiering in World War II (2011)

The American State of Canaan: the peaceful, prosperous juncture of Israel and Palestine as the 51th State of the United States of America (2009)

The Iron Age of Mars: Speculations on a Quantavolution (2009)

THE *QUANTAVOLUTION SERIES (All Metron Publications, 1980-1984 and 2009)*

Chaos and Creation: An Introduction to Quantavolution in Human and Natural History

The Lately Tortured Earth: Exoterrestrial Forces and Quantavolution in the Earth Sciences

Homo Schizo I: Human and Cultural Hologenesis

Homo Schizo II: Human Nature and Behavior

God's Fire: Moses and the Management of Exodus

Solaria Binaria: The Origins and History of the Solar System *(with Prof. Earl R. Milton)*

The Divine Succession: A Science of Gods Old and New

The Burning of Troy: Essays and Notes in Quantavolution

The Disastrous Love-Affair of Moon and Mars

Cosmic Heretics

The Iron Age of Mars: Speculations on Quantavolution (2009)

OTHER WORKS OF ALFRED DE GRAZIA

Michels, Roberto, *First Lectures in Political Sociology*. Translated, with an introduction, by Alfred de Grazia. University of Minnesota Press, Minneapolis, 1949. And Harper & Row, New York, 1965.

Public and Republic: Political Representation in America. Alfred A. Knopf, New York, 1951.

The Elements of Political Science. Series: Borzoi Books in Political Science. Alfred A. Knopf, New York, 1952. And second revised edition: *Politics and Government: the Elements of Political Science*. Vol 1: The Elements of Political Science and Vol. 2: Political Organization, Collier, New York, 1962. And new revised edition: Free Press, New York; Collier Macmillan, London, 1965.

The Western Public: 1952 and beyond. A Study of Political Behaviour in the Western United States. Stanford University Press, Stanford, 1954.

The American Way of Government. National Edition. Wiley, New York (1957). There is also a "National, State and Local edition," publ. by Foundation for Voluntary Welfare.

Grass Roots Private Welfare:: Winning Essays of the 1956 National Awards Competition of the Foundation for Voluntary Welfare. Alfred de Grazia, editor. New York University Press, New York, 1957.

The American Behavioral Scientist, Magazine, founded by Alfred de Grazia, Metron Publications, Princeton, N.J., 1959; acquired by SAGE Press, 1965.

American Welfare. New York University Press, New York, 1961 (with Ted Gurr).

World politics: a study in international relations. Series: College Outline Series, Barnes & Noble, New York, 1962.

Apportionment and Representative Government. Series: Books That Matter. Praeger, New York, c1963

Essay on Apportionment and Representative Government. American Enterprise Institute, Washington,1963.

Revolution in Teaching: New Theory, Technology, and Curricula. With an introduction by Jerome Bruner. Bantam Books, New York, 1964. (Editor, with David A. Sohn).

Universal Reference System. *Political Science, Government, and Public Policy: an annotated and intensively indexed compilation of significant books, pamphlets, and articles, selected and processed by the Universal Reference System.* Prepared under the direction of Alfred De Grazia, general editor, Carl E. Martinson, managing editor, and John B. Simeone, consultant. Princeton Research Pub. Co., Princeton N.J. 1965–69. *Plus* nine more volumes on the subjects of: *International Affairs; Economic Regulation; Public Policy and the Management of Science; Administrative Management; Comparative Government and Cultures; Legislative Process; Bibliography of Bibliographies in Political Science, Government and Public Policy; Current Events and Problems of Modern Society; Public Opinion, Mass Behavior and Political Psychology; Law, Jurisprudence and Judicial Process.*

Republic in crisis: Congress against the Executive Force. Federal Legal Publications, New York, 1965.

Political Behavior. Series: Elements of political science; New, revised edition. Free press paperback, New York, 1966.

Congress, The First Branch of Government, editor, Doubleday – Anchor Books, New York, 1967.

Congress and the Presidency: Their Roles in Modern Times, with Arthur M. Schlesinger, American Enterprise Institute for Public Policy Research, Washington, 1967.

Passage of the Year, Poetry, Quiddity Press, Metron Publications, Princeton, N.J., 1967.

The Behavioral Sciences: Essays in honor of George A. Lundberg, editor, Behavioral Research Council, Great Barrington, Mass., 1968.

Old Government, New People: Readings for the New politics, et al., Scott, Foresman, Glenview, Ill., 1971.

Politics for Better or Worse, Scott, Foresman, Glenview, Ill., 1973.

Eight Branches of Government: American Government Today, w. Eric Weise, Collegiate Pub., 1975.

Eight Bads – Eight Goods: The American Contradictions, Doubleday – Anchor Books, New York, 1975.

Supporting Art and Culture: 1001 Questions on Policy, Lieber-Atherton, New York, 1979.

A Cloud Over Bhopal: Causes, Consequences, and Constructive Solutions, Kalos Foundation for the India-America Committee for the Bhopal Victims, Popular Prakashan, Bombay, 1985.

The Babe, Child of Boom and Bust in Old Chicago, umbilicus mundi, Quiddity Press, Metron Publications, Princeton, N.J., 1992.

The Student: at Chicago in Hutchin's Hey-day, Quiddity Press, Metron Publications, Princeton N.J., 1991.

The Taste of War: Soldiering in World War II, Quiddity Press, Metron Publications, Princeton, N.J., 1992 – 2012 (Revised)

Twentieth Century Fire-Sale, Poetry, Quiddity Press, Metron Publications, Princeton, N.J., 1996.

INDEX

""

transitional forms, 87
tribe, 23, 140, 170, 179, 185

U

V

W

www.ingramcontent.com/pod-product-compliance
Lightning Source LLC
Chambersburg PA
CBHW030006290326
41934CB00005B/247